ROM MODERNIZATION TO MODES OF PRODUCTION

FROM MODERNIZATION TO MODES OF PRODUCTION

A Critique of the Sociologies of Development and Underdevelopment

John G. Taylor

First published 1979 by
THE MACMILLAN PRESS LTD
London and Basingstoke
Associated companies in Delhi
Dublin Hong Kong Johannesburg Lagos
Melbourne New York Singapore Tokyo

Printed in Great Britain
by Unwin Brothers Limited
Old Woking Surrey

British Library Cataloguing in Publication Data

Taylor, John G.
From modernization to modes of production
1. Underdeveloped areas – Economic conditions
2. Economic development – Social aspects
I. Title
301.5'1'091724 HC59.7

ISBN 0–333–24448–6
ISBN 0–333–24449–4 Pbk

Contents

Acknowledgements

I would like to thank John Cowley, Stefan Feuchtwang, and Leslie Sklair for their helpful comments on the first draft of this text. I am also grateful to Elaine Capizzi, David Macey and Doreen Massey for their general theoretical comments and criticisms. My thanks, also, to Abilio Araujo, Antonio Cavarinho, José Ramos-Horta and other Fretilin comrades for their important theoretical points and political inspiration. I retain sole responsibility for the conclusions contained in the text.

April 1978 J. G. T.

Introduction

Sociological analysis of Third World societies is currently in a state of theoretical impasse.

Partial critiques of the Sociology of Development, combined with an increasing awareness of its inadequacies when applied to particular case studies, have led to a virtual rejection of its structural-functionalist tenets. Yet, as a theory, it has not been subjected to a systematic critique. Part-accepted, part-rejected, it lingers on in the hope that, for certain areas of investigation, it is still relevant.

Meanwhile, Underdevelopment theory has been unable to establish itself as a viable theoretical alternative. The vagueness of many of its central concepts has restricted its use. Yet, as with Development theory, critiques have not been directed at the basic concepts operative in its discourse, assessing to what extent they can provide a framework for analysing the structure and reproduction of Third World societies.

Faced with the limitations of these two competing theories, research is increasingly adopting perspectives such as 'Dependency' which can only provide general guidelines for investigation, rather than a theorisation of the reality we are attempting to explain.

Given this situation, the text of *From Modernization to Modes of Production* has a twofold objective: the first part is devoted to a systematic critique of the Sociologies of Development and Underdevelopment, posing questions to their basic notions, analysing to what extent—given the concepts operative in their respective discourses—they *can* provide a basis for explaining the structure, reproduction and future possible development of Third World societies.

Concluding that these concepts are inadequate for analysing crucial aspects of these societies, the second part of the text attempts to develop a new theoretical framework, utilising recent advances made in the theory of modes of production. Third World societies are analysed as particular combinations of different modes of production, which establish a basis for forms of class structure and political representation that are specific to these societies. The framework put forward attempts to provide a means for analysing the non-capitalist societies that preceded colonialism, the effects of different forms of capitalist penetration on them, and the consequent emergence of a form of capitalist development particular to these societies.

Developing the basis for this framework from within the discourse of historical materialism proved to be extremely difficult. The further one investigated the theoretical possibilities, the more questions were raised. For example: where Marx expounds the theory of the elements of modes of production in *Capital*, he does so with only the slightest reference to non-capitalist modes of production. Conversely, in texts whose theoretical adequacy for founding a theory of non-capitalist modes is limited (the *Grundrisse*, the *German Ideology*), Marx constantly raises the issue of these modes, but in a descriptive manner. Furthermore, despite the limitations of these texts, most theorists analysing non-capitalist societies either take their conclusions as fully-formed concepts which can simply be 'applied' to reality, or operate readings of the concept of modes of production which restrict its relevance for our area of investigation. Consequently, I had to undertake a critique of these readings. To add to these difficulties, Marx never analyses the notion of a combination of modes of production, a concept that is crucial for analysing the development of Third World societies. Similarly, whilst historical materialist analyses have described, in general terms, the effects of different forms of capitalist penetration of non-capitalist societies, they have never adequately theorised the determinants of these forms, nor their specific effects on the reproduction of non-capitalist modes of production. These are just a few of the many absences in historical materialist work in this area.

Consequently, rather than being able to put forward an elaborated theoretical framework, it became a question of establishing *a basis for this framework*. As a result, the argument often takes place at a general level of abstraction, and may at times appear somewhat disjointed—exposition being interspersed with digressions into the validity of the concepts utilised. The text may also appear to be sparse on empirical examples. Where necessary, for purposes of clarification or conceptualisation, examples are introduced, or indicated in footnotes—as in the chapter on the Asiatic mode of production, or in the final chapter, where the conclusions are used to briefly analyse development strategies in various countries at different periods. The overall object, however, renders any general discussion of empirical raw material secondary. Given that the key aspect is an elaboration of a new theoretical framework, this could not have been otherwise.

The ordering of the exposition is self-evident from the contents. The critique of the Sociologies of Development and Underdevelopment in Part I is directed specifically at the basic notions of their respective discourses, analysing how the definitions of their object of investigation and the concepts they use in this necessarily limit their explanatory validity, no matter where they are applied. The analysis of the unfounded axioms on which the Sociology of Development is established enables the introduction of the notion of 'restricted and uneven development', as a characterisation of the form of capitalist development currently taking place in Third World countries. Later in the text, this form of development is analysed as resulting from the combination of modes of production and divisions of labour specific to the social formations of the Third World.

The pre-requisites for developing a historical materialist framework are initially stated in the form of theses at the beginning of Part II. The object of study is defined as a social formation in transition from dominance by a non-capitalist to dominance by a capitalist mode of production, as being necessarily structured by an articulation of different modes of production and/or divisions of labour. Given this object, a number of issues are raised. What do we understand by the concept, social formation? If the social formations of the Third World are in a state of transition from one mode of production to another, what effects does this have on the social formation as a whole? How can these effects be analysed? How did a capitalist mode of production emerge to co-exist with a non-capitalist mode in a social formation dominated by the latter? How do we theorise the structure and reproduction of non-capitalist modes of production? What role did the different forms of capitalist penetration play in the historical emergence of the capitalist mode and its articulation with the non-capitalist mode of production? How do we analyse the social structure of Third World formations, given their determination in the last instance by an articulation of modes of production?

Through attempting to answer these problems, a general framework is developed, utilising and extending existing concepts of historical materialism. The final chapter shows how this framework can be used as a basis for analysing economic phenomena, the class structure, and political representation.

One point needs to be stressed. The analysis is concerned specifically with those social formations of the Third World which are undergoing a transition to dominance by a form of restricted and uneven capitalist development, whose reproduction is dependent upon an effective domination of imperialist penetration in various sectors of its economic structure. The limits of our object of study are given by the continuing reproduction of the determinants of the transitional period—that is, by an articulation of capitalist with non-capitalist modes or divisions of labour. Once these determinants no longer structure the social formation, the transitional period can be regarded as being terminated. It is our contention that, within the countries of the Third World, with the exception of a small number of trading entrepot states (for whom the analysis would hardly be applicable anyway), no social formation has yet arrived at this point. However, within the Third World, there are, of course, other transitional formations—those in transition from dominance by an articulation of capitalist and non-capitalist modes to dominance by a socialist mode of production. These formations do not form part of our object in this text; rather, their analysis requires a very different theoretical framework from that which we have elaborated. This may be an obvious point, but it needs clarifying here in order to avoid any misconceptions that might initially confuse one's approach to the text.

The aim of *From Modernization to Modes of Production* is to put forward a tentative general framework for the analysis of Third World formations, which will be further developed by utilising it in an analysis of one particular Third World society during the post-war period. This will be the object of a subsequent text. The extent to which this framework can be successfully developed ultimately to generate the possibility for concrete analyses of particular situations in specific Third World formations, remains to be seen. Whatever the outcome, the procedure adopted in the text, of attempting to elaborate a theoretical alternative, seemed to me to be one way out of the theoretical impasse in which our analyses are currently situated.

Part I
Modernisation and Underdevelopment

1 The Sociology of Development: Theoretical Inadequacies

Although many criticisms have been made of the Sociology of Development, these have either focused on the work of particular theorists or on the limitations of its more dubious notions. There has, as yet, been no attempt to provide an overall critique of its theoretical framework, to show why its conclusions are necessarily limited due to the basic concepts operative in its discourse. Thus, whilst critics have, for example, outlined the errors of notions such as Levy's 'relatively less modernised societies', of Eisenstadt's 'modernisation', of Smelser's concept of 'differentiation', or of Bellah's and Hoselitz's use of Parsons' 'pattern variables', they have failed to specify the theoretical foundations of such notions. Yet, unless we can analyse why the inadequacies of such concepts result from a set of limited notions being continually reproduced in varying conceptual forms, we cannot have any rigorous basis for assessing the explanatory limitations of the Sociology of Development as a whole. In the present situation, where the analytical field of the structure and development of Third World societies is dominated by a theoretical conflict between the discourses of 'Development' and 'Underdevelopment', it is vital that we direct our criticisms at the level of their basic concepts. Rather than assessing the validity of particular texts, we should be asking to what extent these theories *can* generate a framework in which the structure and development of Third World societies can be rigorously analysed.

Consequently, in this first chapter, I have focused on what seem to me to be the basic elements of the discourse of the Sociology of Development, in order to reveal its inherent explanatory limitations.

THE PARSONIAN FOUNDATION

The Sociology of Development is firmly founded upon the theoretical discourse of Parsons. It has relied heavily upon extracting key concepts from this discourse—most notably the pattern variables of action, the functional prerequisites of systems of action, and the functionalist concepts of differentiation and change—and its analysis of the social structure of Third World societies is unashamedly Parsonian.

For this reason, it is essential that we briefly evaluate the theoretical limitations of Parsonian structural-functionalism before going on to examine the other major components that have combined with it to form the discourse of the Sociology of Development. In specifying these limitations, we immediately come up against a difficult problem, since most existing critiques are either partial or inadequate. Rather than directing themselves to the basic elements of the Parsonian discourse and formulating an epistemological critique, they have raised issues which are both secondary—in that they can be shown to be effects or results of axioms which remain uncriticised—and misplaced, since their objections can be adequately met within structural-functionalist theory.

Take, for example, the most popular objections to Parsonian theory in the late fifties and sixties:

The various 'conflict theorists' (notably, Dahrendorf,[1] Rex,[2] and Lockwood[3]) argued that structural-functionalism was incapable of analysing the emergence of social conflict, and, more particularly, that form of disruptive conflict which produced fundamental changes in the functioning of the social system. Yet, if we refer to sections of *The Social System* that analyse social change,[4] or to Parsons and Smelser's formulations in *Economy and Society*, or Smelser's analysis of the Industrial Revolution in *Social Change and the Industrial Revolution*, it is clear that they do provide an account of conflict arising as a result of changes produced in the social structure by transformations in the operation of its various sub-systems. They also provide an analysis of the effects of this conflict on the continuing reproduction of the structure, showing how this necessitates fundamental changes in its reproduction. Because the various sub-systems must fulfil particular functions in order that society can reproduce itself, and because, in order to do this, there must be constant interchanges between systems which develop unevenly, there is a constant basis for the generation of conflict. Contrary to the conflict critics' conclusions, it is the very relations which they criticise between sub-systems and structural reproduction that generate conflict. Furthermore, the 'disruptive' conflict which the critics view as a crucial omission, can also be analysed from within structural-functionalist theory. The uneven development of sub-systems and their elements can generate strains which then provide the basis for a generalised conflict which can only be resolved by major transformations in the dominant value-system or economic structure. Such disruptive conflict, which induces fundamental changes in the reproduction of the structure and its elements, is, for example, the object of Smelser's analysis in *Social Change*. The criticisms of the conflict theorists can, it seems, be met from within Parsonian theory.

Similarly, the criticisms made by writers, such as Barrington Moore,[5] Foss[6] and C. Wright Mills,[7] that the centrality of the concept of equilibrium to Parsonian theory produces an inherently conservative approach, in which order and stability are held to exist empirically at the outset of structural-functionalist analysis, are equally misplaced. In the *Structure of Social Action*, Parsons repeatedly criticises the notion of a simple unitary correspondence between theoretical concepts, facts and actual phenomena. As with all his concepts, the notion of equilibrium has an essentially heuristic status; it is not a readily observable phenomenon, but rather a means of structuring reality for the purposes of theoretical investigation.[8] Far from taking societal equilibrium as a given, Parsons' project is to examine how, given the conflicts perpetually engendered in and by social action and interaction, societies can reproduce themselves whilst retaining relative stability. Concepts such as equilibrium are theoretical means for posing such a question. Whilst Parsons' use of the concept does, at times, blur the epistemological distinction he has established between phenomena and concepts that provide a theoretical knowledge of these phenomena, it does seem that this criticism can again be adequately answered from within structural-functionalist theory. By focusing on Parsons' use of equilibrium in particular analyses,[9] and by abstracting the concept from its context, the critics have largely misunderstood its heuristic status. Consequently, the conclusion that Parsonian theory, by its very content, must necessarily produce a static and conservative bias, due to one of its major concepts, equilibrium, is clearly inadequate.

Again, the set of criticisms centring around Parsons' mystificatory scholasticism,[10] and the inability of structural-functionalism to generate testable hypotheses,[11] or that the theory cannot provide an adequate explanation of the material basis for the legitimate exercise of political power, are equally piecemeal. Rather than examining the basic concepts of structural-functionalist theory, they concentrate on the effects of these concepts.

Finally, with the exception of A. D. Smith's recent work, *The Concept of Social Change*, criticisms of the functionalist scenario of disturbance-adaption-reintegration as a conceptual schema for examining social change have simply denied that the theory *can* explain change rather than focusing on why—in terms of the concepts operative within the theory—the analysis of change is necessarily limited.

Given the partial and limited nature of existing critiques, the first part of this chapter will attempt to provide a more fundamental critique of the overall explanatory adequacy of Parsons' theoretical framework. After outlining the major aspects of the Parsonian discourse, we will examine the epistemological status of the concepts he introduces in his analysis at the action level, analysing the problems generated by these concepts for theorising the mutual determination of action by structure and assessing the concept of structure that ultimately determines action in the Parsonian framework. This description will be followed by a brief formalisation of the Parsonian concept of structure, in order to specify the basic elements of Parsons' structural-functionalist discourse.[12] This specification will enable us to demarcate the major explanatory inadequacies of the Parsonian framework, before going on to an analysis of the structural-functionalist theory of social change.

Given that Parsons' theoretical system has recently been given a much clearer and more precise exposition than it ever received in Parsons' writings themselves,[13] there seems to be little point in providing any such exposition here.[14] Referring the reader to these texts we can, therefore, move directly into a delineation of the basic concepts operative in Parsonian discourse.

THE MAJOR ASPECTS OF THE PARSONIAN DISCOURSE

'Actor' and 'action'

The 'action frame of reference' is Parsons' ultimate reference point. The voluntaristic theory of action that he traces in the writings of nineteenth-century theorists has, as its basic elements, a subject (the actor) placed in an action situation, where he has to organise his behaviour, the immediate means available to him and his interaction with others for the specific means of attaining a particular end or 'goal', which may be either empirical or non-empirical. The action situation is governed both by symbols—through which the actor relates to the various elements of the situation and attributes meaning to them—and by norms, values, rules, etc., which guide the actor's orientation.

This particular conception of the action-situation provides the *ultimate reference point* for Parsons' theoretical system. The subject-actor (which can be either an individual or group) rationally directs its behaviour towards the attainment of a specific goal through the cognitive appraisal of the means available for the attainment of this goal. All action is interpreted from the actor's subjective perspective on the situation: the actor can choose his own goals, the means to attain them and the organisation of these means. In this sense, the theory is voluntaristic—choices only being limited by the constraints of the physical environment and the cultural system, which set limits on available goals and means.

What, then, is the epistemological status of this action frame of reference?

In the *Structure of Social Action*, Parsons specifies the basis on which his notions of action and unit act are formulated:

> These concepts correspond not to concrete phenomena, but to elements in them which are analytically separable from other elements . . . As opposed to the fiction (viz ideal-typical) view it is maintained that at least some of the general concepts of science are not fictional but adequately 'grasp' aspects of the objective external world.[15]

This process of concept-formation, which Parsons terms analytical realism, is held to be a concise description of the process by which scientific thought operates on empirically given raw material. As such, it is governed by an empiricist concept of knowledge. In this conception—as is well known—empirical reality is held to contain elements which are essential for analysing the phenomenon under investigation, and those which are not. Knowledge is already contained in reality as one of its parts, and the task of the theorising subject is to separate or extract this knowledge from the real, by eliminating the inessential aspects of the phenomenon from its essential aspects. Possession of this essential aspect by the theorising subject then provides a basis for the theorisation of the phenomenon under question.[16]

Parsons' epistemological conception of the process by which his concepts are formed appears to be a variant of this empiricist conception of knowledge. The notions of subject-actor, action situation, and orientation of action towards the realisation of specific goals are all formed by extracting essential aspects from a reality which is empirically given. The notion of the subject acting rationally to organise available means to attain subjectively evaluated desirable goals, is a result of the theorist extracting what he considers to be an essential aspect of reality from what is empirically given for him as raw material for his theorising.

Such a conception has severe limitations, which have profound effects both on Parsons' notion of the actor and on his analysis of how this actor orientates himself to his action-situation.

As with all empiricist conceptions of knowledge, Parsons' analytical realism refuses to draw the distinction between what we can term the real object existing externally to and independently of theoretical discourse, and the object of knowledge or thought-object produced as a result of theoretical work operating on existing knowledge. The two are utterly distinct. Whether we are using ideas, notions, or concepts, these can never be considered a simple reflection or direct representation of reality. Knowledge is not contained in reality, but is structured for presentation to the theorist by the combination and inter-relation of ideologies—practico-social, political, religious, theoretical, etc.—that are dominant within the social formation in which he exists. Theoretical analysis always operates on notions embodied in discourses or texts which—however they are transformed theoretically—are originally a product of the reproduction of a particular set of ideologies dominant during the process of theorisation. Furthermore, both the subject *and* its object of investigation are determined by this pre-existing ideological over-determination. Consequently, the crucial question then becomes to what degree do Parsons' concepts remain within the confines of this received frame work, or, alternatively, to what degree do they provide us with a theory that can adequately explain the origins, determinants and continuing reproduction of this framework? Do Parsons' basic concepts of the actor, and his orientation to action tend to reflect, in an abstract form, prevalent ideological conceptions—as does his 'theory of knowledge'—or do they provide us with a basis for analysing the major determinants of these received conceptions?

In much the same way that neo-classical economics (and Marshall, in particular, on whom Parsons relies heavily in the *Structure* . . .) founded its analysis on a perspective which related the organisation of the economy to fulfilling the economic requirements or needs of human subjects, which they took to be empirically given, so too does Parsons found his theory on the supposed needs of individuals to achieve goals through the organisation of available means. Any facts included in his analysis of the social system have their origin in this assumption of the universal need of human subjects:

> *The starting point, both historical and logical*, is the conception of intrinsic rationality of action. This involves the fundamental elements

of 'ends', 'means' and 'conditions' of rational action and the norm of the intrinsic means-end relationship.[17]

Consequently, Parsonian theory can only conceive individual action and social interaction on the condition that it accepts an anthropological conception, which founds all the acts involved in social relations on the rational pursuit of goals by individuals to meet their social needs. Beginning with this assumption, Parsons then approaches 'reality', in order to discover a particular set of given phenomena that can be thought—in one way or another—as effects of the needs of individuals to rationally orientate their action. Thus, the anthropological assumption structures the investigation of reality, and the results of this investigation then reinforce the assumption by enabling concepts to emerge which can develop it in particular theoretical directions (the unit act, the specification of means, forms of orientation to action, etc.). A key question with regard to Parsons' actor is, therefore, the *epistemological status of his assumption of universal need.*

The notion of economic rationality, or that of economic man assumed as given by neo-classical economics and its various antecedents has been fully criticised elsewhere. What it is essential to indicate here, however, is that in the various bodies of economic theory that found this concept,[18] the notion of economic rationality (and the various concepts and laws of operation of the economic structure that are based on it—utility, optimum, equilibrium, marginal productivity, etc.) are, in one way or another, circumscribed from within these theories, either by limiting factors or by their area of application. In the discourse of neo-classical economics, the notion is never conceived as an assumption with universal applicability; the consumer or producer rationally organising his economic existence to meet his economic needs is a means for investigating the operation of the economy whose application to reality is restricted by a series of limiting factors, specified within the theory. Yet, for Parsonian theory, this is not the case. Parsons has abstracted this notion of rational organisation of means to achieve ends from the discourse of neo-classical economics (particularly from Marshall and Pareto) and, discovering it in Durkheim and Weber, he has then elevated it to the status of a concept with universal application, gradually discarding the limits placed on its application by neo-classical economic theory.

The consequences of this for the Parsonian conception of the actor are considerable. The notion of rational action—the basis of the voluntaristic theory of action—becomes a concept that can be universally

generalised only because it approximates extremely closely to common-sense notions prevalent in the practico-social ideologies dominant in the industrial capitalist social formation in which the theory emerged. Consequently, its status as a theoretical concept is highly questionable. Whilst there are acts which we can classify in these terms, there are many for whom the description is quite inapplicable. The action of an owner of an industrial firm in reorganising his operations on the factory floor to increase output and achieve a higher rate of profit is a rational act in the Parsonian sense, but how could we possibly classify social acts such as smoking a cigarette, taking a walk, reading a book, and so on, as being acts in which the actor has one particular explicit end-in-view during his action, and is organising available means for achieving this goal? In this sense, all action obviously cannot be classified as rational. Yet, if we now introduce into this schema the notion that all action must in some way or another meet the requirements that are determined for it by the social role in which it takes place, since the reproductive needs of the social structure require that such roles be played, and no matter whether actors are conscious of it or not, in this sense they must always direct their actions, we *can* conclude that all actions are goal-directed. Having done this, however, we have generalised the concept of rationality to such an extent that the initial meaning of the actor constantly having in view an explicit goal *cannot be established by voluntaristic action itself*. The conclusion, therefore, can only be meaningful if we introduce into the unit act arguments concerning the functional needs of the social structure, with the result that despite his own subjective intentions the actor's action has to be rational.

Consequently, Parsons' notion of rationalism in itself establishes very little. It only has any meaning when it is inserted into arguments concerning the reproductive needs of the social structure. The notion of 'goal-directed behaviour' is not, therefore, a founding concept produced through theoretical work on given raw material, but rather an empirical reflection of existing conceptions of what it entails to be human, as represented in prevailing ideological notions whose essential aspects Parsons has extracted.

The actor's 'choice': the pattern variables

This conclusion, viewing Parsons' work on the actor, orientation, and unit act as a classification of common-sense, can be further exemplified by examining his basic conception, in the *Social System*, that a 'scientific' classification of the relational aspect (i.e. of the forms governing interaction between actors in a unit act) can be produced by using the terms 'cognitive', 'cathectic', and 'evaluative', which, according to Parsons, constitute 'the three basic modes of motivational orientation', 'empirical clusterings' of the 'most elementary components of action in general'.[19] Each of these 'scientific classifications' can be replaced by 'common-sense' axioms: 'Cathectic'—one must consider what can be gained from any action, and at what cost to oneself, before one embarks on it. 'Cognitive'—one must consider how the action can be successfully carried out. 'Evaluative'—one needs to choose the most appropriate means of success. Basically—in the case of any act—how does one feel about something and how can one think of the best way of attaining it in a given situation? The fact that these concepts—whose objective is to provide a classification through which we can begin to theorise *why* interaction takes different forms in different unit-acts—can be so easily transformed into common-sense statements of the most general kind, illustrates again how, in formulating his basic action concepts, Parsons compounds an invalid empirical extension of concepts (such as rational action) with a reproduction of existing ideological notions to produce his theoretical schema.

Parsons' approach is, however, best exemplified by his pattern variables—the mutually exclusive and universally applicable set of choices (or 'dilemmas of choice') open to actors in orientating their action to achieve their goals within unit acts—a set that is deduced from the cognitive/cathectic/evaluative scheme outlined above.

Parsons claims that this set of alternative choices establishes 'a system covering all the fundamental alternatives which can arise directly out of the frame of reference for the theory of action.'[20]

As Rocher indicates,[21] there is nothing particularly orginial in these formulations: one side of the set corresponds to Töennies formulation of 'gemeinschaftlich' and the other to his 'gesellschaftlich' relationships as outlined in *Community and Association*, where they exist as relations produced by different types of social structure. Abstracted from this theoretical context, they are used by Parsons as a means for descriptively classifying action. As such, they again tend to reproduce a number of commonly accepted ways in which individuals conceive of their relations to each other, rather than producing any knowledge of why this should be the case.

Take, for example, the dichotomies 'self-orientation/collectivity-orientation' and 'affectivity/affective-neutrality'. Both can be transposed into the common-sense choices of one's attitude being determined by self-interest or not, and its embodiment in action requiring emotional expression or not. In this categorisation, Parsons appears, therefore, to be listing features which characterise social action and interaction, which is presumed to be goal-oriented. The categorisation is nothing more than a limited classification of the ways in which individuals ideologically conceive of their inter-relations with other individuals, or their attitudes towards the goals which (according to Parsons) their activity must inherently be directed to. The concepts are a simple reproduction of these notions in another form.

But beyond this, one can also pose the 'pattern-variables' the question—in what sense *do* individuals 'make choices' before they act? Parsons provides us with a definite answer to this, of course. He argues that the choices can be explicit or implicit;[22] that is, the actor can be consciously aware that he is choosing, or unconscious of the fact of choice. This point illustrates the fundamental problem with the pattern-variables. In addition to their proximity to common-sense notions and their non-exclusiveness, they cannot, in themselves, explain anything, unless we refer them to a further postulate, namely that, whether or not the actor is 'aware' of it, he must necessarily make a 'choice', because the reproductive requirements of the social structure require that he do so.

It would appear, therefore, that, as with the concept of rational actor, the variables 'orientating actors to action' can only go beyond empiricist reflections on common-sense notions and have some explanatory power if they are referred to a classical functionalist notion of the social structure which requires that action take such a form.[23] The argument for the actor's unconscious choice has no more validity than an unproven assumption in Parsons' analysis of social action until he relates it to a postulate concerning the needs of the social system.

Thus, the basic concepts of the voluntaristic theory of action, formed through the empiricist process of analytical realism, reproduce and categorise prevalent ideological notions of behaviour and social relations in a theoretical discourse which can only establish the 'universality' of the concepts as a result of a pre-given notion of the prerequisites for the functioning of the social structure. In themselves such notions as the pattern variables do nothing more than classify a number of aspects of received patterns of behaviour. Their universality cannot be established by such classification for the reasons given above.

Consequently, within Parsons' discourse, we see a continual displacement away from voluntaristic to functionalist postulates, away from an empiricist categorisation whose explanatory power is necessarily minimal, to the formulation of a structure whose needs require the existence of the very 'orientations to action' that Parsons initially tried to establish as universals (in *The Structure of Social Action* and *The Social System*).

This transition from the voluntaristic to the structural level is, however, achieved through an *assumption of functionalist axioms* concerning the reproduction of the social structure. Consequently, it is these assumptions, and not any voluntaristic postulates that are ultimately determinant in Parsonian theory. The common-sense notions of action and action-orientation are only given a theoretical status and explanatory power by the introduction of these assumptions. This is illustrated very clearly at a number of points in Parsons' texts, where these assumptions are introduced to provide a means of moving from the action to the structural levels, *in the absence of any adequate theorisation of this transition except in terms of the presumed needs of the social system*. Such points are sections III and IV of the *Social System*, and, more clearly, the article, 'Pattern-Variables Revisited: A Response to Robert Dubin',[24] with its object of 'generalising the integration of pattern variables into the four-function paradigm', an article which outlines the organisation of the social structure as an effect of its functional prerequisites. It is necessary, therefore, to briefly examine these two texts.

Action and structure

In the *Social System*, the pattern variables, extracted for heuristic purposes from the actual real process of interaction, are combined. Their logical possible forms of combination are established, with the object of setting up 'a classification of value-pattern types defining role-orientations'. Such a process is, according to Parsons, 'the *fundamental starting point* for a classification of possible types of social structure.'[25] According to this formulation, then, a classification of the variables governing orientation to action *can* form the basis for a theory of structure.

Having reached this point, where particular aspects of action and interaction have—through 'analytical realism'—been extracted from the empirically given to construct a theory of action, Parsons then attempts a 'proof' for the validity of his combinations of action. In order to prove that the combinations he has specified can provide us with a basis for classifying social structures, Parsons returns us, however, to the very reality from which the pattern-variables were initially extracted. In order to show that his combinations are not simply 'randomly related', but organise action and interaction, Parsons proves this by showing how 'in reality' combinations of pattern variables 'actually' cluster together empirically in particular systems of roles (kinship, stratification, etc.).

The tautology of this proof is immediately apparent.

The validity of the pattern-variables, which are no more than descriptions of aspects of the empirically given, is established by reference to their existence in that given. Such a process fails to provide us with any basis for establishing *why* particular 'empirical clusterings' exist as sub-systems of roles, since the explanation can necessarily be nothing more than a description of action.

Consequently, Parsons is forced to go beyond this level if he wishes to show how the pattern-variables can be a 'fundamental starting point for classifying *types of social structure*'. At this point in his argument, he must, therefore, make the transition between the action and the structural levels, if his theory is to transcend a simple description of what exists.

To this end, he poses the question (in Ch. V. of the *Social System*) of the relation between the empirical clusterings of action and the 'functional imperatives' of the social structure. How, he asks, can we relate the existence of orientations to action and their institutional framework to the reproductive requirements of the structure, to 'the conditions which must be met by any social system of a stable and durable character'?[26]

This question of how such a relation can be theoretically formulated, is posed very clearly at this point. Yet, symptomatically, rather than being directly answered, it is substituted for another, very different question. Namely, *assuming that order and stability must exist*, since the interactive relations between ego and alter 'must' be regulated, how can the various sub-systems (kinship, etc.) co-exist with, or be adapted to each other, and what other structures must exist for the integration between sub-systems to be realised? In other words, to resolve the fundamental problem of the theoretical inter-relation between action and structure, Parsons introduces the *unproven axiom* that *stability already exists.*

Consequently, the problem posed at the outset, of theorising how Parsons' conception of action necessitates and requires the functional imperatives he subsequently outlines, of how his notions of action and structure *mutually determine* each other, slides under a second question, whose assumption requires that a solution to the first question has already been arrived at.

We have, therefore, a fundamental split within Parsons' discourse: on the one hand, a descriptive classification of action, and on the other, a conception of structure whose reproductive needs (the functional imperatives) require sets of pattern-variables organised in sub-systems. The mutual determination of these two aspects is nowhere theorised. They are simply tied together through the introduction of an unproven assumption; this assumption then determines the appearance of the functional imperatives, and it is only through the introduction of the latter that the existence of clusterings of pattern-variables can be provided with a theoretical determination that goes beyond description. Consequently, for Parsons, the only means of transcending the limitations of his descriptive notions of action is through the introduction of functionalist postulates which are not derivable from these notions.

This conclusion is also evident in the two texts, *Towards a General Theory of Action* and *Economy and Society*, but here the problem itself is never posed: action is analysed solely in terms of the reproductive requirements of the structures' functional imperatives. We will return to this when we outline the basic elements of the Parsonian discourse, but, for the moment, we must briefly examine the one other point in Parsons' discourse where the mutual determination of voluntaristic action and functional needs is posed, but then displaced to the functional level through the introduction of an unproven assumption.

The object of Parsons' article, 'The Pattern Variables Re-visited', is to 'generalise the integration of pattern variables into the four-function paradigm'; that is, to theorise an admitted absence in his work.

He begins by describing the pattern-variables that cluster within the roles required for the carrying out of two of the structure's functional prerequisites, 'pattern-maintenance' (Pm) and 'goal-attainment' (Ga). He then argues that, for these pattern-variables which govern interaction to exist, it is necessary that,

> In such a structured system both actor and object share institutionalised norms, conformity with which is a condition for stability of the system. The relation between the actor's orientations and the modalities of objects *cannot be* random [our stress].[27]

Consequently,

> It is *in the nature of an action system* to be subject to a plurality of functional exigencies [our stress].[28]

Because the norm of 'stability' for interaction is assumed in this way, the Pm and Ga subsets then necessitate further sets of pattern-variables which can ensure the 'equilibrium' of the system. These additional subsets are the integrative (I)—the means through which 'the elementary actor object units' are integrated and the adaptive (A)—'the mechanism by which the system as a whole is *adapted to the environment* within which it operates.'[29]

The argument therefore is clear. Two major clusterings of pattern-variables can be derived from a classification of interaction itself. For these clusterings to be reproduced, they require the existence of further clusterings or combinations of the five dichotomies of choice for action. Certain subjects are, therefore, derived from a classification of action, whilst others are derived from the functional requirements of a structure containing the initial subset. Herein lies the 'resolution' to the problem of the mutual determination of voluntaristic and functionalist perspectives on the structure.

Yet throughout the analysis, it is, of course, the functional level that ultimately exercises determination. The Pm and Ga sets of pattern-variables are not, despite Parsons' claims, established solely at the descriptive level; rather they cluster precisely because, as Parsons has already indicated in *Towards a General Theory of Action*, they are required to do so by the structure's major functional imperatives, Pm and Ga. Furthermore, even if we accept that they *can be* deduced solely from a descriptive classification at the action level, it remains that Parsons never indicates how the functional exigencies can in any way be derived from this classification. Rather, he simply introduces the unproven assumption of stability and equilibrium. Having introduced it, he can then bring in the functionalist theory of structure, since this is 'required' by the assertion itself. Consequently, not only are the Pm and Ga sets ultimately determined by the functionalist conception of structure, but so also are the I and A clusterings which can only exist given the presumed need by systems of action for institutionalised stability.

It appears, therefore, that, despite Parsons' attempts to formulate a framework in which the *mutual determination* of action and structure can be analysed, the gap between action—as a set of description categories—and structure—as a theory of the reproductive prerequisites that govern action—remains the major theoretical vacuum in his discourse. As a result, even in expositions directly concerned with this problem, Parsons' voluntaristic postulates still retain a purely functionalist determinacy. Despite the stated problem of conceptually theorising the inter-relation between pattern-variables and functional imperatives, the 'Response to Dubin' appropriately concludes by referring to 'the normative requirements of its [the acting unit's] action necessitated by the functional exigencies of the system.'[30]

The concept of structure

Given the failure to theorise the mutual determination, and the resultant emergence of the functionalist notion of structure as ultimately determinant in Parsonian theory, it is crucial that we now move on to analyse the basic elements of this notion.

The Parsonian concept of the reproductive requirements or functional imperatives that all societies must fulfil is sufficiently familiar to avoid detailed exposition here. The imperatives of goal-attainment, latency, integration and adaptation constitute functions to be fulfilled by sub-systems within the structure. Each sub-system, in turn, must itself also fulfil the four imperatives, and this necessitates a whole series of interchanges between sub-systems. This framework determines which set of pattern variables co-exist in each sub-system. As Parsons states:

> The conceptual scheme of the four-system paradigm has added a set of rules and procedures whereby the analysis of the components of action can be carried out by 'looking down' on them.[31]

The major question for us here is, rather, on what basis are these functional imperatives which determine patterns of action within the social system actually conceptualised?

We have already outlined how the notions of 'stability' and 'equilibrium' emerge within Parsons' discourse, as assumptions filling a theoretical vacuum. Yet, in addition to providing a means for 'interrelating', 'actor' and 'structure', the unproven assumption of stability also provides a basis on which the notion of the functional imperatives can be founded. For Parsons' argument is, tautologically, that, if a system is functioning and maintaining itself in the way that—given, the above—it has to, then it 'must be' responding to a set of needs which ensure this maintenance. The main problem then becomes one of discovering these 'needs'. On this theoretically unfounded basis, the functional imperatives are constructed.

Assessing the validity of these imperatives returns us to aspects of Parsons' discourse that we criticised earlier in the chapter. We noted that the notion of goal-directed rationality could only transcend the level of common-sense description if it was inserted into a theory of the prerequisites for structural reproduction. Yet, we now find that a major functional imperative of this structure, namely 'goal-attainment', is simply an extension of this notion to the structural level. All societies, by their very nature, must, according to Parsons, establish particular goals and methodologically pursue them. The acceptance of this rationalist postulate, at the structural level, is absolutely crucial, since it is this capacity of action systems which distinguishes them from 'non-action' systems, such as the biological or physical.

All our earlier criticisms of goal-directed behaviour as a means for analysing action equally apply, however, to this notion of 'systemic goal-attainment'. Indeed, the very notion that is supposed to provide a basis for transcending the limitations of rational action by situating it within a theory of the social system that goes beyond the level of description, is, in effect nothing more than a generalised re-affirmation of the initial description, unprovenly extended to the level of a systemic need, on the basis of a tautological assumption.

The same process is at work in the 'integration' imperative. The social system must not only rationally direct itself to the achievement of goals, but it must also attain the consummatory end of integration, organising unit acts for this purpose. This second imperative has, therefore, a dual foundation—as a corollary of the assumed need for stability for interaction to persist, and as a generalisation from the notion of rational action, since, on the basis of the first assumption, all societies must 'necessarily' achieve the goal of integration.

Once these notions of rationally directed behaviour and the need for integration have been unprovenly introduced at the structural level, then two further imperatives, adaptation and latency, follow as corollaries, since they are nothing more than necessary means whereby the systemic ends of goal-attainment and integration can be achieved. For the social system to achieve these goals, it must utilise resources available to it from the environment in which it exists. The environment is constituted by the other action systems (the biological, cultural and personality), which similarly utilise means available to them from the social system. In much the same way, the latency (or pattern-maintenance) imperative exists as a further means for achieving systemic ends; for goals to be attained, a set of unit-acts must be created which supply actors with the motivation necessary for their attainment. Thus, the function of the latency imperative, transmitting the symbolic representations, values, ideas, etc. of the cultural system into the social system, has its foundation solely in the Parsonian notion of systemic rationality.

It appears, therefore, that Parsons' theory of the social structure, of the determinacy of action by the need to meet a set of functional imperatives, is formed on a twofold basis: on an unfounded generalisation from a descriptive notion of action, and on an unproven assumption introduced to fill a crucial absence in Parsons' discourse. Neither of these provides us with a basis for transcending the limitations inherent in his initial concept of rational orientation to action, since neither has any validity beyond an acceptance of this notion. In place of the stated object of producing a theory of the social structure which has an explanatory power above and beyond the acceptance of this descriptive axiom, we are simply given an unfounded extension of this generalisation as a substitute for theorisation. The failure to conceptualise the mutual determination of action and structure is now compounded by a simple structural determination based on two notions—the one limited by being confined to description, and the other an unproven assumption.

Having reached this point, we can now formalise Parsons' conception of structure which lies at the centre of his discourse—a centre whose basis remains descriptive and untheorised.

A formalisation of the concept of structure

The object of Parsonian analysis is, in each case, a more or less persistent phenomenon ('the family', 'the economy', 'social stratification', etc.) within the social system. The task of the analysis is to show that the social system is in an internal state of stability or equilibrium, and in an environment with certain external conditions based on utilisation of resources from other action systems, such that, under these internal and external conditions, the phenomenon analysed has effects which fulfil some functional requirement of the social system. In so doing, the phenomenon thus contributes to meeting a set of conditions (the functional imperatives) necessary for the system remaining in a stable working order.

Thus, a Parsonian explanation of a particular phenomenon would take the following form: At a particular moment in time, the social system is in a state of equilibrium, this being determined by internal conditions (the organisation of sets of pattern variables and exchanges between them) as required by the operation of the functional imperatives combined with external conditions (the utilisation of resources available from other action systems). It is then argued that the social system can only achieve and maintain this state of equilibrium if a particular condition—such as supplying actors with necessary motivation ('latency')—is satisfied. This is followed by the contention that, if a phenomenon such as socialisation and the mechanism through which it is achieved, the family, *were* present within the social system, then, as an effect, the condition of latency would be satisfied, thereby contributing to the state of equilibrium. The conclusion then follows that, at this point in time, the phenomenon of the family functioning to socialise its members to the dominant values of society in order to produce a state of equilibrium, is, in fact, present within the social system. The phenomenon of the family is subsequently analysed from within this particular perspective.

From this explanation, we can draw the following conclusions: For Parsons, the contents of the social structure, namely the pattern-variable sets existing in particular phenomenal forms, are unified and inter-related through being realisations of functions delineated as essential for the stability of the structure. As aspects of the structure, they are equivalent, since they are all phenomena through which these basic functions are fulfilled. The essence of the structure, which determines the appearance of the pattern-variable sets is, therefore, the general notion of function, introduced through the assumption of equilibrium and specified in the descriptively-based functional imperatives. This essence requires an empirical realisation of itself in particular combinations of pattern-variables. Consequently, the latter can only exist as phenomena, and can only be inter-related in particular ways through the action of the essential functional essence of the structure. This essence is immediately legible in these phenomena, and in the relationships (boundary exchanges) necessitated between them. The conception of the social structure is, therefore, governed by a specific form of causality which we can term an 'expressive' causality.

> It [the concept of expression] presupposes in principle that the whole in question be reducible to an *inner essence*, of which the elements of the whole are no more than the phenomenal forms of expression, the inner principle of the essence being present at each point in the whole, such that at each moment it is possible to write the immediately adequate equation: *such and such an element* (economic, legal, literary, religious, etc.) = *the inner essence of the whole*.[32]

Once Parsons' particular conception of the totality of the social structure has been specified in this way, and once the mode of causality this necessitates has been demarcated—i.e. once the major elements of the discourse have been laid bare—it seems to me that we now have a more adequate basis for delineating the major limitations of Parsons'discourse as necessary effects of his conception of structure.

THE MAJOR INADEQUACIES OF THE STRUCTURAL-FUNCTIONALIST DISCOURSE

Returning to our Parsonian explanation above, its most striking absence is revealed in the final element of its exposition, prior to its conclusion—namely, from the statement, 'that if a particular phenomenon is present in the social system, equilibrium would be attained' the conclusion is deduced that the phenomenon *is* present in the social system.

As a number of authors, notably Hempel,[33] have pointed out from an analysis of functionalist explanations in general, such a conclusion could, of course, only be inferred if the argument asserted that *only* the presence of this phenomenon could satisfy the conditions (i.e. the meeting of a particular functional prerequisite) required for equilibrium. Yet, Parsons never claims this for the phenomena he analyses. Indeed, in *Essays in Sociological Theory*, for example, he assumes the existence of what he terms 'functional equivalents', phenomena that fulfil the same function(s) as he is attributing to the phenomenon under investigation.[34] Consequently, it could be the case in Parsons' analysis that the existence of any number of alternative phenomena known or unknown could produce, just as the phenomenon under investigation, the conditions necessary for the state of equilibrium. Given this, his explanation cannot explain *why the particular phenomenon analysed* rather than any one of its equivalents *is actually present* in the social system at the moment of investigation.

In addition, assume for the moment that Parsons *could* establish the existence of a set of known functional equivalents *sufficient for the existence of equilibrium*. Even if his analysis could attain this level, the most he would be able to conclude would be that of a class of phenomena (although we cannot definitely know precisely which), some are present in the social system at the moment of investigation. The analysis, therefore, *fails to provide us with any basis for explaining the necessity of a particular phenomenon* as a form of combination of a specific set of pattern-variables, since its presence within the system cannot be deduced from the functional requirements of the social system, as Parsons outlines them. A major limitation of the Parsonian discourse is, therefore, that the analysis undertaken of a particular phenomenon cannot in itself establish adequate explanatory grounds for expecting the occurence of a set of pattern-variables rather than one of its 'functional equivalents' in any given social system. Consequently, the predictive capacity of the theory is extremely limited. Given that a phenomenon's existence cannot be deduced from the structure whose imperatives Parsons claims determine it, then the presence of the phenomenon can only be established on two bases: first, on an *ex-post-facto* basis ('since a particular phenomenon fulfilled a certain function in another structure, it is likely to operate similarly in the structure under investigation' etc.) and, secondly, through a process of descriptive classification of the empirically given (establishing whether or not pattern-variables 'actually do' cluster in reality, etc.), in which case the theory of the functional imperatives governing the appearance of pattern variables is redundant. In either case, a major limitation of the discourse remains unresolved, and the extent to which one can generalise or predict the existence of a phenomenon and its function in the reproduction of the social structure remains extremely limited, being confined to generalisation on an *ex-post-facto* empirical basis, whose relevance to the structure under investigation can in no way be guaranteed.

To introduce us to a second major inadequacy in Parsons' discourse, let us assume, for the moment, that one can, in fact, specify a *mutually exclusive set* of functional imperatives or an *indispensable function* applicable to a particular combination of pattern-variables and appearing in a phenomenal form in different social systems. Even if this could be shown to exist, it could only be given adequate explanatory and predictive power if the internal and external conditions determining its appearance could be clearly specified. This leads us on to a second problem which concerns more directly the predictive status of Parsonian analysis.

The Parsonian explanation effectively makes an inference from the existence of a particular functional imperative in a given social system to the assertion that, in another social system this imperative will be met in a similar way. Such an argument, however, presupposes a general principle, namely that, within certain limits of adaptability to its environment, a social system will—by developing appropriate phenomena—satisfy the various functional requirements produced by changes in its internal state and in its environment. However, this, of course, presupposes a theory of the self-regulation of social systems—that within a given range of conditions, of the social system *vis-à-vis* its environment, and of its organisation internally, the functional imperatives will continue to be fulfilled. Unless such conditions are specified, we have no basis for concluding that social systems will continue to meet these imperatives at a future time. Yet, Parsonian theory has singularly failed to produce a theory of the conditions under which the functional imperatives will continue to be fulfilled. This is most notable in the accounts of the relation of the social system to its so-called environment (viz. the other action systems); rather than formulating a set of external conditions whose reproduction ensures that the social system *can* continue to regulate itself to its environment (and vice-versa) in meeting the functional imperatives, Parsons is mainly concerned with how a social system can utilise resources provided by the other systems of action in order to achieve the functional imperatives, *on the assumption that equilibrium already exists and generates a 'need' for the functional imperatives to be met*. In *Societies*, for example, delineating the relations between the social system and its physical environment, Parsons states:

> Resources must be allocated toward the satisfaction of the vast variety of wants present in *any* society . . .[35]

Where the analysis of the relations between the social and the other systems of action *is* carried beyond this point (for example, in *Societies*, pp. 10–18), it achieves no more than a descriptive classification of basic mutual requirements:

> Thus a society's *primary* exigency vis-à-vis the personalities of its members is the motivation of their participation, including their compliance with the demands of its normative order.[36]

Similarly, if, within the social system, we pose the question of the range of pattern variable sets beyond which the functional imperatives cease to have a basis for operation, this again remains unresolved in Parsons' discourse.

The results of this second limitation are extremely damaging for Parsonian theory. Since it cannot establish the conditions under which a social system regulates itself—internally and in relation to its environment—to provide a basis for the emergence of the functional imperatives, and since this gap in the discourse is simply filled by the tautological assumption that equilibrium must be attained in order that such systems can survive, we have no basis either for arguing that these functional imperatives will be met or that the phenomenal forms through which they are exercised will exist to fulfil these functions in other social systems, present or future. Consequently, the Parsonian theory of the social structure and its functional prerequisites gives us no adequate basis for predicting the future possible development or state of a social system or systems. To its failure to explain the occurrence of phenomena within the structure on the basis of its notion of the functional imperatives, must now be added its failure to predict either the recurrence of these phenomena or their functional imperative determinants in social systems other than that from which these phenomena and imperatives are initially deduced.

Finally, and perhaps the most seriously for Parsons' discourse, his two fundamental concepts, 'equilibrium' and the 'functional imperatives' are mutually determined through a tautology whose elements are untheorised. On the one hand, the state of equilibrium can only be ensured through the 'adequate' operation of pattern-variable sets structured by the needs of the functional imperatives, and, on the other, for these pattern-variable sets to exist and inter-relate adequately, there must be a state of equilibrium. Furthermore, as we have seen, of the two mutually determining notions of this tautology, the one has the status of an unproven assumption introduced into the discourse to fill a fundamental yet unresolved problem, and the other is nothing more than an unfounded extension of an inadequate description of action to the systemic level. This unfounded tautology at the very centre of the Parsonian discourse, further limits the already limited predictive and explanatory capacity of the theory.

From our specification of the Parsonian discourse and the mode of causality operative in it, we can, therefore, draw the following conclusions:

1. The possibility of explanatory generalisation between social systems is severely limited, since *the presence and function of any phenomenon within the structure can only ultimately be established at the level of descriptive generalisation.* This is so because the theory of the social system cannot provide us with an adequate explanation of the existence of any particular phenomenon as an effect of the structure and its reproduction.
2. The combination of this necessarily descriptive generalisation with the inability to theorise the internal or external conditions for the operation of the functional imperatives, and thereby the process of self-regulation of the system, means that *the theory's capacity for prediction is confined to ex-post-facto empirical description.* A phenomenon is analysed as embodying a particular pattern-variable set for the fulfilment of a specific function solely because it was analysed as such in a pre-existing structure. Similarly, the functional needs (imperatives) of the system can only be established as generalisations made from the assumed requirements of a past structure.

3. In addition to these explanatory and predictive limitations, the two notions governing the *ex-post-facto* and descriptive generalisations that can still be made are themselves established on the basis of an untheorised tautology.

These three fundamental inadequacies of Parsons' discourse, uncovered through a specification of its basic elements, already indicate that any attempt to found a theory of social change from within Parsonian confines must necessarily be seriously limited. The restrictions placed on explanatory generalisation, predictability, and on an analysis of causal relations that does not lapse into a tautological framework are such that they must necessarily generate an inadequate explanation of change.

Yet, despite these limitations, it was such a theory, generated from within these confines, that combined with Parsons' theory of the social structure to lay the foundation for the Sociology of Development. Having analysed the limitations of Parsons' concept of the social system, we can now turn to this theory of change, showing how it both perpetuates and extends the fundamental inadequacies of Parsons' discourse.

THE STRUCTURAL-FUNCTIONALIST THEORY OF SOCIAL CHANGE

Here again the major postulates of this theory are sufficiently known to avoid repetition, and since I am largely concerned with specifying the theory's limitations, the outline that follows will necessarily be brief.

As Parsons states in his article, 'A Paradigm for the Analysis of Social Systems and Change', the theory of change takes as its initial basis the assumption of equilibrium:

> The concept of equilibrium is a fundamental reference point for analysing the process by which a system either comes to terms with the exigencies imposed by a *changing* environment, without essential change in its own structure, or fails to come to terms and undergoes other processes, such as structural change . . .[37]

At any one point in time, the social system is analysed as being in a state where its sub-systems are adequately integrated and the functional imperatives are being adequately met. The need for change in such a system is produced either by the emergence of strains[38] in the relations within or between sub-systems, or by new factors introduced into the system from other social systems, and from the cultural, personality and organism systems of action that co-exist with the social system. Changes upsetting the original equilibrium have an 'endogenous' or 'exogenous' origin, or may be a combination of these. We thus have an inital state of equilibrium subject to a 'disturbance' which then upsets the existing relations within and between sub-systems. Examples from functionalist literature abound: Smelser's analysis of the disturbing impact of technological innovations originating in the economic sub-systems on this and other sub-systems leading to the emergence of the industrial revolution; Bellah's analysis of the impact of Western ideologies on the cultural system of 'traditional' societies, and so on. In each case, the emergence or introduction of a qualitatively original phenomenon into a particular sub-system requires change in the processes of goal-attainment, integration, adaptation and latency that govern the sub-system's reproduction, and this, in turn, affects the input-output relations which this sub-system has with other sub-systems, and—above all else—begins to prevent the system from attaining its goals. Consequently, the system must now move from its initial equilibrium point to a new equilibrium, in which the changes required by the introduction of the new element have been successfully incorporated into the system. Yet, in making this move—and here we come to a fundamental concept in functionalist analyses of change—the social system must create new roles in which the new tasks required by the incorporation of the new element can be carried out, and it must also create new norms legitimising these new roles. Consequently, the original, more simply organised system must, as a result of this incorporation, become more complex. Hence, as a corollary of induced transition from one equilibrium level to another, society must become more differentiated both in its roles and in the cultural legitimation of these roles in the value system. This process of 'structural differentiation' is a necessary corollary of disturbance to equilibrium; as societies change, they must necessarily become more differentiated. Smith, in his text, *The Concept of Social Change*, gives a succinct characterisation of this notion:

> Everywhere we find more complex social organisation growing out of the failure of simpler structures to fulfill their functions effectively. Everywhere we witness the breakup of less specialised units, unable to meet new problems and conditions, and their replacement by two or more specialised units which perform these functions and meet these new tasks with greater efficiency.[39]

For structural-functionalism, increasing differentiation is a postulate which has universal validity, since all transitions from one level of equilibrium to another necessarily require further specialisation.

'Reintegration', the other major concept of the theory of change, follows as a direct corollary from this postulate. Since the period of transition between equilibrium levels necessarily involves such features as disjunctions between the new roles required and existing patterns of behaviour, conflicts between values appropriate to different states of equilibrium, and—most importantly—a need for the new more specialised roles to be fully integrated into the system during the tranisition period, it follows that reintegrationary mechanisms must be created to re-equilibrate the system at its new level and to handle conflicts that emerge during the transition perid. This entails either the transformation of existing organisations, or the creation of new institutions specifically to regulate new forms of conflict. Examples of the former are Eisenstadt's 'modernising élites' promoting greater political participation to institutionalise conflicts produced in the 'modernising' societies of the Third World, and, of the latter, Smelser's notion of the creation of the trade union to regulate conflicts generated by the newly emerging capitalist division of labour in the late nineteenth-century British economy.

We have, therefore, a basic schema: disturbances in an equilibrated system, generated by a combination of endogenous and exogenous factors, necessarily produce increasing structural differentiation, which, by definition, requires re-integrative mechanisms in order that a new level of equilibrium can be attained. Yet, behind these processes—according to functionalist theory—there also lies an implicit objective, whose attainment governs the whole movement and direction of change: that a system previously 'adapted to its environment' (the surrounding action systems) must ultimately become 're-adapted' at a new level of functioning; and since, at this new level, the system is, according to functionalist postulates, more differentiated and can perform more specialised tasks than previously, it follows that society must be more efficiently adapted to its environment than in the pre-existing state of equilibrium. As Parsons states:

> If differentiation is to yield a balanced, more evolved system, each newly differentiated sub-structure must have increased adaptative capacity for performing its *primary* function, as compared to the performance of *that* function in the previous, more diffuse structure.[40]

Thus, the whole direction of development of differentiation and reintegration is determined by a presumed need for the social system to adapt itself to its environment. The impact of the disturbing element is analysed in terms of its enhancing the adaptive capacity of the social system via the process of differentiation; the need for society to be more adequately adapted to its environment, which is indicated by the emergence or introduction of the initial 'disturbance' to equilibrium, is met through differentiation which must, by definition, fulfil this need. Thus, behind the processes of differentiation and re-integration lies an all-pervasive need for adaptation—a need which appears to be both the determinant and the result of the transition from one state of equilibrium to another.

Having briefly outlined the functionalist schema, we can now examine its major tenets, focusing on how they reproduce and extend the limitations of the Parsonian discourse as we have outlined it.

THE MAJOR TENETS OF THE FUNCTIONALIST THEORY OF CHANGE

Causation as given

The first point to note about the schema is that the exogenously or endogenously developing phenomena that produce the initial disturbance remain untheorised. Since the focus of the theory is on the *response of the system* to changes in its equilibrium state, it must take the theorisation of this state on the basis of the unproven assumption of equilibrium as its starting point. Generally, therefore, the disturbing phenomenon is taken as a given, its causes remaining untheorised; when functionalist texts go beyond this, they simply indicate that the disturbance is a result of previous differentiation in other systems. Parsons' analyses of the transitions from one societal type to another (in *Societies*, for example) are particularly clear examples of the failure to theorise the phenomena that disturb equilibrium. In each case, the sources of change of historical transitions are traced to elements which are introduced into the analysis as givens, unrelated to the actual process of adaptation increasing differentiation—which Parsons is supposedly analysing.

We have, therefore, a remarkable contradiction in the functionalist analysis of change. Despite its claim to provide a theoretical framework for explaining change, the phenomena that produce change in its schema remain untheorised. Rather, they are simply accepted as empirically given. Since the theory is concerned with the response of the system to change, rather than theorising the causes of change, one must then ask, what relevance has it for explaining the determinants of change beyond a simple acceptance of the inter-relations between events as they were empirically presented in the ideologies in which these events were actually lived? It would appear that, by its necessary focus on the initial equilibrium state, functionalist theory is unable to theorise the very phenomena that induced the transition it is claiming to analyse; all it can do is describe these phenomena.

Ex-post-facto generalisation

Secondly, since the transition from one level of equilibrium to another is determined by the need to incorporate the effects of an intervening phenomenon within sub-systems whose organisation and inter-relation are determined by the need to meet the functional imperatives, and since this overall framework provides no theoretical basis for establishing the existence of any phenomena in the social system beyond the level of descriptive generalisation, then it follows that the conclusions derived from analysing the effects of social change on phenomena in any particular system are not in themselves *theoretically valid for other systems*. The explanation of social change in terms of adjustments to equilibrium to meet the needs of functional imperatives cannot, therefore, *in itself*, establish what qualitatively new phenomena will emerge within the structure, or which phenomena will be transformed in what direction, since these are not derivable from the theory itself.

Functionalists are, therefore, necessarily forced into a position of *ex-post-facto descriptive generalisation* from transitional periods *as they have previously occurred.*

Differentiation as an 'explanation' of change

This inability of structural-functionalist theory to generate explanations of change and directions of change without having to revert to a pre-theoretical level of descriptive generalisation, has led functionalists to focus more on the concept of differentiation, since they initially viewed this as a notion which could 'resolve' these problems. As a result, differentiation has become both a means of classifying societies and, ultimately, an 'explanation' of change.

This is most apparent in Parsons' text, *Societies . . .*, when he classifies social systems on a universally applicable traditional-modern continuum on the basis of the degree of differentiation in their economic, political and cultural levels. Beyond this, however, he also views differentiation as a *cause* of social change, as *a set of structural preconditions necessary for* the attainment of modernisation. In *Sociological Theory and Modern Society*, for example, he declares that only societies with advanced levels of differentiation can become industrialised.[41] The effect, or the result of the transition from one level of equilibrium to another becomes, therefore, a *cause of change itself*. Such a conclusion is also found, for example, in Smelser's models of social change.[42] Yet, it is clear that structural differentiation as a concept cannot have such explanatory validity. Far from providing an 'explanation' of change, all it does is describe, in the most general terms possible, the effects of industrialisation on the structure of those societies *which have already experienced industrialisation within a specifically capitalist framework*. Clearly, they have become structurally differentiated, but what value does this *ex-post-facto* description have for explaining the causes and future possible directions of change in the contemporary 'modernising societies' of the Third World, as functionalists term them? The only conclusion that can follow from the functionalist schema is that, in order to industrialise, these societies must differentiate in the direction of the 'end-state' reached by those societies that have already industrialised, since the fact that differentiation is a necessary correlate of industrialisation is 'proven' by the very existence of these societies.

Ex-post-facto description thus becomes a substitute for explaining social change in the functionalist schema. The specification of stages through which societies must necessarily pass is put forward as a solution without any analysis of the causal connection between these stages, or of the determinants of differentiation itself. When this question is posed, the theory—as we have seen—simply introduces contributory factors to change, whose effects are established outside the framework of the theory.

If the theory itself cannot establish the causes of transition from one level of differentiation to another, then as an explanation of change, it is redundant. Since the concept of differentiation fails to do this, and merely describes a general effect of industrialisation as it has previously occurred, it must come within this category.

In addition to this, however, two further points should be noted. First, that the level of generality of the concept of differentiation is such that its relevance to modernisation *per se* is extremely limited. What is required, if we operate within the functionalist scheme, is an analysis of the causes and characteristics of the transition between specific equilibrium states in particular Thirld World societies as they 'modernise' in the contemporary period. Yet, as a number of authors have indicated, differentiation, as a general descriptive trend, can be shown to have been applicable to currently industrialised societies at all stages of their development (even at times when certain sectors were becoming momentarily 'de-differentiated'). What, therefore, one must ask, is its relevance for explaining *specific* periods of change, such as those through which Third World societies are now passing—societies which, as distinct from those on which Parsons' analyses are based, have been subject (as we shall see)—to a particular form of restricted development resulting from penetration by societies industrialising or already industrialised? Taking the concept of differentiation, how do we distinguish between the specific characteristics of this period, as opposed to previous periods of change? The extreme generality of the concept renders the very posing of this question impossible.

Secondly—and this point is crucial—since the functionalist theory of change establishes—through *ex-post-facto* generalisation—an evolutionary correlation between industrialisation and differentiation, it can only resolve the problem of future possible directions of change in Third World societies by referring them to a *particular end-state*, namely that attained by the contemporarily *most differentiated* social systems. On the basis of its evolutionary postulates, a universal historical path towards greater differentiation emerges, which all social systems *must necessarily follow* if they are to industrialise. Hence, any form of social and economic development which does not 'fit' into this schema must necessarily be analysed as a deviation, a short-term failure which must be overcome in order that the system can return to its pre-determined path of ever-increasing differentiation. Consequently, the 'Europocentric' bias evident in functionalist theories of modernisation is not—as some authors have suggested—simply a reflection of the ideological interests of particular theorists, but a *necessary effect* of the theory in which they operate.

It seems, therefore, that the concept of differentiation provides no solution to these inadequacies of the theory of change; rather, it compounds them, by forcing the analysis in directions that cannot cope with the emergence of forms of social structure that are not immediately apparently equivalent to forms that have existed or already exist. As a substitute for theoretical analysis of a changing social system, its determinants and its future possible directions of change, functionalists are reduced to a process of the most general empirical classification of societies, a classification derived solely from trends presented empirically to them by the surface phenomena of capitalist industrialisation as it has occurred historically in those societies which are currently the most highly differentiated. Consequently, social change as it is now taking place in contemporary Third World societies can—by the very nature of the theory itself—only be analysed on the basis of the extent to which it does or does not contribute to the attainment of this particular end-state. Prediction is necessarily confined within this empiricist strait-jacket of classificationary stages.

An untheorised tautology

Finally, we should note that, as with the structural-functionalist theory of the social system, so too is the theory of change based on a tautology whose foundation is untheorised.

In each particular case of change to which it is applied, the functionalist theory of change tacitly assumes that the overall cause of differentiation is given by the need for a unit to adapt to its environment. Yet, this differentiation is then analysed as producing—as a consequence of change—a greater adaptation of the system as a whole to its environment. Within the functionalist schema, we thus have an implicit tautology, in which the need for greater adaptation is both a cause and a consequence of change. The determinant of the process is its necessary result.

Furthermore, if we examine the theoretical foundation for the need for societies, in Parsons' phrase, to produce an 'enhancement of adaptive capacity', we find that its derivation is also empiricised, since it is deduced from a 'universal' trend towards structural differentiation, which—as we have seen—is based on the most general result of industrialisation in already 'differentiated' societies. The very essence of the theory remains within a descriptive generalisation—at the very point from which the construction of a theory should begin if its concepts are to have any claim to theoretical status.

From our examination, it seems that the structural-functionalist theory of change is seriously limited on several levels. Firstly, it provides no adequate theoretical basis for analysing the causes of change. Secondly, it cannot establish any theoretically valid conclusions concerning the effects of change that are generalisable from one social system to another. Finally, it cannot provide any basis for analysing future possible directions of change, except by relating them to an end-state which already exists; consequently, any qualitatively new forms of economic and social development generated in the course of 'modernisation' can only be labelled as 'deviants' from this norm. This seriously restricts the theory's validity for analysing contemporary Third World societies, the development of whose structures—as we shall see below—are clearly distinct from the forms taken by structural differentiation in the industrial capitalist societies from which the functionalist end—state is naïvely extrapolated.

ELEMENTS FOR A THEORY

Despite its various inadequacies, the functionalist theory of change, together with the Parsonian concepts of structure and functional prerequisites constituted the foundation of what came to be known as the 'modernisation theory' of the 'Sociology of Development' as it emerged during the post-war period.

Together with other concepts abstracted from the writings of theorists such as Tönnies and Weber, the functionalist notions of structure and change were eclectically inter-related to produce a theory whose object was to provide a framework in which the transition from traditional to modern forms of society—a process which the social formations of Africa, Asia and Latin America were said to be undertaking—could be analysed.

It is essential, therefore, that we assess the effects of these concepts that were combined with the theories of structure and change.

Tradition–modernity

Since all societies must undergo change from one equilibrium-state to another, it is essential, for the 'modernising' states of the Third World, that we know *from what state to what state* change is proceeding. The introduction of the tradition/modernity couple was an attempt to answer this question by establishing a framework within which the process of differentiation could operate.

The couple itself is, of course, a familiar heritage of nineteenth-century sociological development. Yet, as with other theorists before them, writers such as Hoselitz,[43] who first applied the traditional/modern dichotomy, gave their own particular interpretation of the characteristics of the 'traditional' and 'modern' society. Just as they conceived the dichotomy as a 'unit-idea' that could be extracted from the discourse of differing theorists, so too they regarded the Parsonian set of pattern-variables—conceived theoretically as a limited set of choices presented to the actor as an effect of the structure—as a notion that could be utilised regardless of its theoretical determinants in the structural-functionalist discourse. The argument was amazingly simple.

Whilst one side of the pattern variable choices was held to characterise traditional societies, the other side characterised modern societies. Hoselitz constructed two ideal-types of society, the one combining universalism, functional specificity, achievement-orientation, and collectivity-orientation (the modern type); and the other combining particularism, diffuseness, ascription and self-orientation (the traditional type). Modernisation (to be achieved through a process of increasing differentiation) then became the problem of ensuring a transition from dominance by the traditional to the modern type of orientation of action.

The effects of this conception were clear: separated from their structural determinacy, the pattern-variables became a means for classifying societal types. As opposed to theorising the structure and reproduction of a Third World society in structural-functional terms, one could now situate it *purely in empirical terms*. A society's potentiality for 'modernisation' could be assessed solely in terms of the relative absence or presence of characteristics of action, *regardless of their structural determinants*. This entailed a massive empiricisation: the theorisation of change as an effect of a disturbed equilibrium at different moments in the differentiation of modernising societies was relegated to a secondary position as an empirical search for 'modern'-ideal-type orientations emerged with an increasing focus on modernisation as a product of changes in action-orientation, regardless of their structural determinants—determinants, which, as we have seen, are ultimately the theoretical foundation of the structural-functionalist system.

Since we will be assessing the results of this empiricisation below, I now want to examine a second element which combined with the structural-functionalist discourse to produce modernisation theory.

Rationalisation

Whilst the traditional/modern dichotomy indicated the origins and end-state of the process of differentiation, the introduction of the Weberian notion of rationalisation established a more particular direction for the specific periods of re-adaptation experienced by Third World societies, since an increase in rationalisation came to be a pre-condition for the attainment of modernity.

As a result, the theoretical framework already established became strengthened both teleologically and empirically.

Teleologically: to the conception of modernisation as a continuum with its successive moments linked by the need to achieve a pre-determined end-state, was now added a further notion which similarly read off history in the past anterior, but which could give more particular indicators of the degree of modernity, by specifying the extent to which institutions such as religion, law and politics were characterised by the development of a world outlook based on the rational calculation of means towards goals that could be realised empirically. The degree of 'modernisation' thus became subject to measurement in a highly descriptive manner.

Empirically: on the one hand, rationalisation came to be considered a generator of change, in the sense that an increase in normative patterns of rationality at the cultural level produced a more systematic programme for modernisation in the social system. Yet, on the other, the determination of this rationalisation remained untheorised, being explained as a conjunction of events in much the same way that the 'causes' of the initial disturbance were 'analysed' in the structural-functionalist theory of change.

Modernising agents

Having specified the 'origins' and 'ends' of the process of change, the mechanisms through which this process unfolds, and the specific ways in which the degree of attainment of the ends can be measured, thereby concretising the application of functionalist theory to the modernising societies of the Third World, the Sociology of Development required one final specification: namely the 'agencies' whereby the given end could be attained. By what means could the processes of differentiation, rationalisation, and the dominance of certain pattern-variables over others be ensured?

The answers were manifold. Processes like urbanisation, the strengthening of institutions such as education or islamic religion, the promotion of rural entrepreneurs and traders, the creation of innovating political élites, a more nationalistic military, etc.—all these, it was argued at one point or another, could provide a more adequate basis for Third World societies to attain their immutable action-oriented destiny.

Increasingly the only way in which sociologists of development analysed aspects of these social systems was as a mirror-image of their descriptive evolutionary schema. The theory of modernisation became a search for empirically given indicators of a pre-given teleological conception of history, whose explanatory validity remained unquestioned.

This search was, of course, further empiricised by the introduction of conclusions derived from motivational psychology, conclusions, which regarded the 'potentiality for modernisation' as a psychological attribute of individuals. Since this latter perspective has been fully criticised elsewhere,[44] I intend only briefly to indicate its limitations; my main concern is to show how its appeal for the Sociology of Development is an effect of a continual discarding of theoretical explanation of change in favour of an empirical search for particular action-orientations—an effect produced as a result of the combination of the elements we have examined with the structural-functionalist theory of structure and change.

The impact of psychological reductionism

The approach of the psychologistic school to modernisation can best be approached by quoting its leading exponent, McClelland, in his major text, *The Achieving Society*:

> In its most general terms, the hypothesis states that a society with a generally high level of need achievement will produce more energetic entrepreneurs who, in turn, produce more rapid economic development.[45]

Development, therefore, becomes solely a problem of ensuring that those individuals who possess a high motivation to achieve, a drive that is as innate as, for example, the hunger drive, enter into key entrepreneurial roles. Once this is guaranteed, economic growth is ensured. Whilst, at particular moments in his work, McClelland hypothesises a causal relationship between 'individual mysticism as a dominant religious ethic', 'early independence training' and high 'n achievement', these remain purely at the level of correlations (and even then they are inadequately established), and the focus is constantly on the need for individual transformation, creating 'n achievers' through such means as Protestant conversion, education, re-organising fantasy life or decreasing father dominance. Thus, for example, in an article devoted to analysing 'Motivational Patterns in S.E. Asia',[46] he concludes that the Chinese government has been able to achieve a more rapid economic development than the Indian government solely because Chinese society contains more individuals with a high level of 'n achievement'. Leaving aside the many inconsistencies and limitations that have been cited in McClelland's proof of the existence of his achievement drive, both from within motivational and non-motivational psychology,[47] the limitations of his overall approach to the problems of modernisation are manifestly obvious.

No matter how many 'n achieving' individuals are capable of being recruited into entrepreneurial roles, this can produce no effect unless changes within the economic structure can create such roles. Even if we accept that the form of economic individuality specified by McClelland is important for Third World modernisation (which, of course, assumes that this development can only occur on a capitalist basis) and his analysis of it, it remains the case that the emergence of such individuality has its own structural determinants. The calculating entrepreneur, organising production rationally to achieve the highest rate of profit possible under given conditions of capital and labour-power can only emerge when the economic structure provides a basis for constantly generating and reproducing these conditions. The latter, in turn, require political and social guarantees for their continuing reproduction, as we shall constantly see throughout our analysis. Whether one regards the 'traditional' social structure as the major barrier to the creation of these conditions (the Sociology of Development), or one analyses the continuing impact of the advanced industrial economies as restricting their emergence (the Sociology of Underdevelopment), there can be no doubt that the major problem is a structural one. As is shown, in fact, by McClelland's own programme in India, the inculcation of an 'entrepreneurial ethic' can provide no basis for generating domestic economic growth when the conditions for its realisation—namely the enlarged reproduction of a growing industrial capitalist sector geared towards the domestic market—do not exist. The problem, therefore, is precisely to locate the *structural reasons* for this limited capitalist development, rather than attempting to impose a form of individuality whose insertion into the economic sector is necessarily restricted.

THE EFFECTS OF THE NEW COMBINATION

Having reached this point, we can now assess the effects of the combination of the theoretical aspects outlined above with the structural-functionalist theories of structure and change. Do they provide a basis for overcoming the limitations inherent to those theories, or do they result in their reinforcement? The introduction of the traditional and modern ideal-types has a number of important effects.

First, it produces a tendency to move from a structural-functionalist theorisation, both of the social system that pre-existed colonial penetration, and of the effects of this penetration on that structure—of the degree to which it introduced disturbances and induced differentiation and greater adaptation—to a concern with *describing the effects* of such change. If this process of change induces a move from dominance by a traditional to a modern set of pattern-variables, what we want to know, for any one structure, is the degree to which such changes have facilitated this move, and to what extent further changes may further facilitate it. Yet, we can only know this by analysing the determinants of this changing dominance, of the effects of changes in the structure on action within it; and once we define traditional and modern societies primarily in terms of sets of variables of action, then the problem of their structural determinants has no need to be posed. How do we explain the existence of particular combinations of pattern-variables at any moment in time, or analyse their future possible development, if we have abandoned the structure that—as we have seen—necessarily determines them? The introduction of the traditional/modern type as an action-oriented pattern-variable dichotomy takes the theory further in the direction of empirical description, reinforcing the split that already exists within it between action and structure, thereby focusing increasingly on a *description of action* rather than a *theorisation of its determinants*.

Secondly, the way in which the two polar opposites, traditional and modern are conceptualised as ideal-types produces considerable problems. As writers with such diverse theoretical perspectives as Bendix[48] and Frank[49] have indicated, societies which sociologists of development refer to as archetypical of the traditional or modern form can be shown to exhibit elements of both action-sets. Furthermore, elements from either side of the dichotomy can be shown to be crucial for the reproduction of both the traditional and modern forms.[50] Consequently, the heuristic status of the concept is extremely limited in ideal-type terms; the theory has failed to produce a dichotomy that can meaningfully differentiate between societal forms that are crucial to it—in that they demarcate the origins and end-state of the change that the theory is supposedly facilitating.

Given the absence of an adequate theorisation of the two 'end-points' of the theory of change, the stage of transition then becomes purely a matter of empirical preference. Whilst some authors stress the presence of particular action-sets, others stress the prevalence of other sets—with no theoretical means of seriously assessing the validity of their respective conclusions for the future dominance of one set over another. This is particularly evident, for example, in Levy's notion of 'relatively less modernised' and 'relatively modernised' societies;[51] two volumes are devoted to descriptively classifying societal types, in the absence of any theorisation of how their structural development has determined, determines or will determine the relative dominance or subordination of one set of pattern-variables over another. The end-result is that the explanation of the state of transition comes to depend even more heavily on *ex-post-facto* generalisation; the level of modernisation of a society is established by comparing it historically in a totally empirical manner with similar states previously attained by contemporary industrial societies in the course of their specifically capitalist development. The theory is necessarily forced into this position by the nature of the traditional/modern conceptual dichotomy introduced into it.

Turning to the results of the insertion of the concept of rationalisation, we can similarly deduce a series of theoretical effects.

The substitution of description for theorisation in the evolutionary teleology, as evidenced in the traditional/modern dichotomy, is reinforced here by the need to read off the present levels of rationality in relation to a given 'rationalised' end; the problem again becomes one of *empirical assessment* on the basis of an *ex-post-facto* analogy with the norms of market calculation historically required by capitalist industrialisation. As we indicated earlier, the teleological framework is reinforced by the introduction of rationalisation, and within this framework, the explanation of directions of change is reinforced by a description of the *effects of change* at different moments in the transition to modernity.

One of the major problems with the structural-functionalist theory of change is its inadequate basis for theorising the causes of change. This led it to focus on the process of change (differentiation) to the detriment of its determinants, which were ultimately factorally accepted in the form that they were historically given. Similarly, the introduction of rationalisation requires a focus on the historical realisation of a phenomenon, rationality, states of whose realisation need no more determinacy than the need for a rational essence to be progressively developed in the interests of the end-state. Thus, as with the notion of differentiation, the determinants of an intensification of rationalisation appear outside the scope of the theory's object. Consequently, the problem of empiricising the causes of change, endemic to the structural-functionalist theory of change, is further reinforced by the introduction of rationalisation.

The notion of modernising agents marks the culmination of the theoretical effects of the polar ideal-types and the notion of rationalisation. In search of these agents, the Sociology of Development is forced to focus on particular aspects of the social structure, describing these on the basis of unproven assumptions (the end-states), and increasingly failing to explain theoretically their determinants or their future possible directions of change in relation to the structure in which they exist. From theoretical explanation in a teleological framework whose limits can be meaningfully established, to empirical description in a teleological framework whose limits cannot be meaningfully set—such seems to be the overall direction produced by the combination of these 'new' elements with the structural-functionalist theory of structure and change.

CONCLUSION

In this chapter, we have tried to show how, as a result of the basic concepts operative in its discourse, the Sociology of Development is necessarily restricted in the very areas—of causation, comparison and prediction—which are crucial for analysing Third World societies. No matter what textual form its concepts assume, these restrictions render the theory inadequate for analysing its object. As a result of our analysis, we now have a more rigorous basis for assessing the theoretical validity of the Sociology of Development, and a terrain on which the limitations of particular texts can be more systematically approached.

Beyond this however there is, of course, a further more fundamental problem which we have ignored throughout this chapter. The Sociology of Development accepts as an unfounded axiom that the industrialisation of Third World countries must necessarily follow a path analogous to pre-existing forms of capitalist industrialisation. It also accepts as axiomatic that the processes of economic penetration of Third World economies which have characterised both the colonial and neo-colonial periods, are essential prerequisites for modernisation. It is this assumption which, of course, the Sociology of Underdevelopment rejects, arguing its converse, that an ending to these forms of penetration is an essential prerequisite for the development of an economic structure based upon the needs of the indigenous population.

Consequently, having examined the limitations inherent in the discourse of the Sociology of Development, I now want briefly to examine the validity of these axioms, before going on to an analysis of the Sociology of Underdevelopment. This will be done by indicating descriptively the major characteristics of the economic relations contemporarily existing between the industrial capitalist economies of Western Europe and the U.S. (defined by modernisation theory as approximating most closely to the ideal-type modernised end-state) and those economies of Africa, Asia and Latin America (defined by modernisation theory as in transition to the 'end-state'), whose reproduction depends on a continuing intervention by the industrial capitalist economies. Outlining these relations will enable us to assess the extent to which one can establish an equivalence between previous forms of capitalist industrialisation and the patterns of development characteristic of contemporary Third World societies.

2 The Sociology of Development—Unfounded Axioms: The Restricted and Uneven Development of Third World Economies

In the preceding chapter, we outlined the major limitations of the Sociology of Development by assessing the adequacies of its concepts in analysing its theoretical object—the so-called transition of Third World societies from 'tradition' to 'modernity'. To these limitations must now be added what—as several authors[1] have indicated—is its crucial error, the (often unstated) assumption that underlies its theoretical foundation.

For the Sociology of Development, the process of 'modernisation' now occurring in Third World societies must follow a path similar to that taken by the industrialisation of the capitalist formations of Europe and North America. Whether we are faced with a descriptive stages theory in which Third World countries are placed on a continuum of economic development derived from previous phases of capitalist industrialisation, and their level of transition from 'tradition' to 'modernity' measured by various indices, or with a quasi-Weberian formula which searches for the universal cause ('rationalisation') of capitalist industrialisation, hidden and suppressed in the interstices of the social structure, the assumption remains the same. As in the work of Parsons, so in that of the other major theorists, the task is to discover the fundamental traits 'responsible for' capitalist industrialisation, argue that they are universally necessary for 'modernisation' *per se*, and then examine how they can be diffused into contemporary Third World formations.[2]

Many authors—particularly the 'Underdevelopment' theorists[3]—have provided a critique of this assumption. They have attempted to show in general terms how the process of capitalist industrialisation in each of its specific phases was facilitated by the enlarged reproduction of the capitalist mode of production occuring within a world division of labour[4] which guaranteed that the reproductive requirements of each phase could be adequately met through an exploitation of non-capitalist and emerging capitalist economies. They have also established that such a guarantee is impossible for contemporarily modernising societies, since the very economic, political and ideological relations that presently exist between the industrialised capitalist societies and those of Asia, Africa and Latin America which remain within the capitalist world division of labour, act to confine and restrict development to particular sectors. Consequently, not only do the same basic economic preconditions for an extension of capitalist industrialisation as occurred in Western Europe not pertain in Third World societies, but the very attainment of these pre-conditions is blocked by their economic relations within the now industrialised capitalist countries. It is for these reasons—as is well known—that theorists such as Frank and Baran argue that they constitute a unified object of study, the 'underdeveloped societies'.

We will be examining the notions of the underdevelopment theorists in the following chapter, but, for the moment, I want to briefly outline in as concise a form as possible both the general aspects of the economic structure of the so-called 'modernising societies' of the Third World and the major characteristics of their economic relations with industrialised capitalist formations. The focus is limited. Our object is simply to establish that the result of economic penetration of Third World economies has been to produce an economic structure whose development is both highly uneven and necessarily restricted to particular sectors by this penetration. We will show how the existence of this uneven and restricted form of development belies any application of conclusions derived from the process of Western European industrialisation, a process whose development was never constantly confined by these limiting forms. The points raised have been dealt with elsewhere by underdevelopment and dependency theorists,[5] but I want to bring them together in a concise form at this point both to criticise a key tenet of the Sociology of Development and to provide an introduction to the main body of the text, with its object of providing a theoretical framework in which the social formations of the Third World can be analysed.

Delineating the major characteristics of different sectors of production, the composition of trade and the nature of commodity exchange between Third World and industrial capitalist economies, we will outline a set of general characteristics which both demarcate the specificity of the economic structure of Third World societies and reveal the effects of their reproductive dependence on the enlarged reproduction of industrial capitalist formations.

THE DEMARCATORY TRAITS OF THIRD WORLD ECONOMIES

The agricultural sector

One of the most striking demarcatory features of the economic structures of Third World societies is the percentage of the population engaged in agriculture. Take, for example, the figures in Table I, showing agricultural population as a percentage of total population.

TABLE I Agricultural Population as a Percentage of Total Population

Continent	*% of population engaged in agriculture 1970*
World	51
N. America[a]	7
Europe	19
Oceania	22
S. America[b]	39
Asia	64
Africa	69

[a] Includes Central America
[b] Excludes Central America
Source: Food & Agricultural Organisation of the UN Yearbook 1974

Despite the fact that such a large proportion of the population is engaged in agriculture, the contribution of the underdeveloped countries to world agricultural output is, however, far less than this proportion would indicate.[6] The level of agricultural productivity is much lower in the Third World than in the industrial capitalist economies. This is indicated in Table II, where I have selected crops that are staples for both underdeveloped and industrial capitalist economies.

TABLE II Yields for selected crops (in Kg per hectare) 1974

Crop	N. America	W. Europe	Africa	Asia	Latin America	Developed[a]	Developing[b]
Wheat	1751	3367	728	1176	1489	2138	1168
Rice	4978	5256	1307	1935	1866	5580	1872
Maize	4476	4180	1108	1046	1392	4008	1280
Barley	1910	3334	968	1190	1130	2691	1059
Potatoes	26555	21998	6123	8367	8933	22747	8848
Ground Nut	2793	2244	664	812	1075	2287	797

[a] Includes Oceania and Other Developed Market Economies
[b] Includes the countries of the Near East and Other Developing Economies
Source: Food and Agricultural Organisation of the UN Yearbook 1974

This gap in productivity levels has remained constant during the post-independence period, since overall food production per capita has not dramatically increased in Third World countries. Indeed, the overall per capita increase for the developed countries has been greater than that for the underdeveloped countries. If anything, therefore, the difference in productivity levels has increased.

If, bearing this per capita increase in mind, we now focus on a comparison of production in the export sector with that in the domestic sector producing crops for indigenous consumption, the contrast is stark. Take, for example, the following figures showing the percentage increases in a number of commodities which are largely produced by Third World countries and consumed by industrial capitalist nations.

Whilst the average increase in output for these export crops during the period 1964–74 was 52.6 per cent, *the amount actually consumed per capita in the domestic sector decreased* during the same period. Consequently, whilst the production of crops for export is increasing, the amount of food produced for local consumption is actually decreasing in per capita terms.

On the one hand, we have a sector characterised by low levels of productivity, primitive techniques, low levels of output, and often a resultant failure to meet the needs of the domestic market, and, on the other, a sector which is the reverse of this situation, but whose growth is founded on an external demand unrelated to the basic requirements of the domestic agricultural economy.

The extent to which agricultural sectors of Third World countries are geared towards the production of cash crops for export is illustrated in such figures as 53 per cent of India's exports being devoted to tea, 60 per

TABLE III Index numbers for food production and per capita food production 1963–74 (1961–65 = 100)

Region		*1963*	*1964*	*1965*	*1966*	*1967*	*1968*	*1969*	*1970*	*1971*	*1972*	*1973*	*1974*
N. America	FP[a]	102	102	105	109	114	115	115	113	124	122	126	128
	PC[b]	102	101	103	105	109	108	107	104	113	110	111	110
W. Europe	FP	101	102	103	105	112	115	115	117	121	119	125	130
	PC	101	101	101	103	109	110	109	110	114	111	116	120
Latin America	FP	100	103	107	108	114	115	120	124	125	127	128	135
	PC	100	100	101	99	102	100	101	102	100	98	97	99
Africa	FP	101	103	106	104	110	114	119	121	124	124	121	127
	PC	102	101	101	97	100	102	103	102	101	99	94	96
Asia	FP	101	104	101	102	107	113	118	124	125	121	132	128
	PC	101	102	96	95	97	100	102	104	102	97	103	97
Developed	FP	102	103	104	108	114	116	116	116	123	122	126	128
	PC												
Developing	FP	104	104	104	105	110	114	119	124	125	125	129	131
	PC	101	101	99	97	100	101	102	103	102	99	99	99

[a] FP = Food Production
[b] PC = Food Production Per Capita
Source: Food and Agricultural Organisation of the UN Yearbook 1974

TABLE IV Percentage increases for selected crops: 1956–64 and 1964–74

Crop	*1956–64*	*1964–74*
Coffee	46%	64%
Cocoa	70%	33.5%
Tea	28%	42%
Natural Rubber	18.5%	65%
Bananas	46%	48.7%

Computed from Jalée (*Third World in World Economy*), p. 27 and from UN Production Yearbooks to 1974

cent of Costa Rica's exports to bananas, 50 per cent of Malaysia's exports to rubber, 65 per cent of Ghana's exports to cocoa, 60 per cent of exports in the Dominican Republic to sugar, etc.[7]

At the same time, many Third World countries are actually importing foodstuffs to meet the needs of their domestic economies. Countries such as Indonesia, Chile and India are dependent on such imports whilst concomitantly much of their most fertile land is being devoted to the production of cash crops for export. Furthermore, this trend is increasing; the imports of food by underdeveloped countries have risen from $4,000 million in 1955 to $8,000 million in 1970.[8]

The degree to which industrial capitalist countries are dependent on the importing of these crops has been adequately stated elsewhere.[9] Here it is sufficient to specify a number of examples.

TABLE V Production of particular crops in Africa, Asia, Latin America relative to their production in the developed capitalist nations (1973) (in metric tons)

Crop	*Third World*	*Developed capitalist nations*	*Third World products as a percentage of world production*
Tea	1,056,885	95,100	91%
Groundnuts	12,635,000	1,950,000	84.6%
Natural Rubber	3,449,369	0	100%
Cocoa beans	1,355,205	0	100%
Coffee	4,185,687	983	99.98%
Jute	2,404,281	270	99.99%

Source: U.N. Production Yearbook 1973

In the most general terms, therefore, a defining trait of the agricultural sector of Third World formations is an extreme unevenness between their domestic and export-oriented sectors. This unevenness is the result of a process of economic dependence for the export sector and a restriction on economic development in the domestic sector. No matter what form was taken by capitalist penetration of the agricultural system, it created a subsistence sector where development was dominated by an export sector dependent on the enlarged reproductive requirements of the industrialist capitalist economies.[10] Whether its objective was to utilise the existing agricultural system to produce export crops by restricting domestic production, or to create an entirely new system of land tenure exclusively for export crops, or forcibly to undermine the existing system in order to create a system of labour-intensive plantations, in each case the most productive resources were increasingly devoted to commodities required for the reproduction of the industrial capitalist economies.

This dependent development had as its corollary a restriction of the domestic sector. With its resources increasingly taken from it, with its markets undercut by the entry of new products, and with no adequate form of assistance given to it, the domestic economy generally remained at or below subsistence level, its plight being expressed most acutely in its (well-published and much misunderstood) 'land/population' ratio—a direct effect of capitalist penetration transferring, under-mining or destroying the indigenous agricultural economy.

Consequently, the restrictions contemporarily placed on the development of indigenous agriculture are a specific result of the creation and continuing reproduction of an internally dominant yet dependent sector by capitalist penetration. As long as this sector remains, the possibilities for utilising the resources of the agricultural economy to ensure the enlarged reproduction of an indigenous economic structure centred on the requirements of the domestic market will continue to be severely restricted.

Thus, the conclusion put forward by the Sociology of Development that a 'modernisation' of the agricultural sector must be based on a replication of Western European agricultural development, relying on a filtering through of techniques developed in the export sector can be seen to be grossly inadequate. The specific forms of dependency, the extremely uneven sectoral configurations, and the restriction of the agricultural sector by the reproductive requirements of a dominant sector with contrary objectives—none of these were the major problems facing the capitalisation of agriculture in the now industrialised capitalist states.

In seeking to discover the mirror-image of this process in the Third World, the Sociology of Development is founded on a denial of the specificity of the latter's contemporary situation, a specificity which supposedly forms its theoretical object.

The industrial sector

This sector is, again, characterised by an extremely uneven configuration.

As with the production of cash crops, the process of *extraction of raw materials for export* plays an absolutely crucial role in many Third World countries; so much so that, for countries such as Chile, Iraq, Iran, etc., the export of one commodity effectively determines and conditions the continuing reproduction of their entire industrial sector.

TABLE VI Raw materials of export of selcted countries as a percentage of their total exports, averaged for 1965–8

Country	*Percentage*	*Raw materials*
Algeria	66.5	Crude petroleum, iron ore, phosphates
Congo-Kinshasa	83.2	Copper, diamonds, zinc ore, tin ore, cobalt
Nigeria	52.1	Petroleum, oil seeds
Zambia	95.7	Copper, non-terrous metals
Bolivia	82.2	Tin ore, lead ore, silver, zinc ore, wolframite
Chile	79.2	Copper, iron ore, various minerals
Venezuela	93.1	Crude oil, refined petroleum products, iron ore
Iran	87.9	Crude oil, petroleum products
Iraq	94.2	Crude oil

Sources: Jalée, 1969, Statistical appendix and various U.N. Yearbooks

The major direction of these exports is revealed by the table that follows; it shows that, for many essential products, Third World economies remain a privileged source of supply, and that their share of world production remains considerable. Indeed, they remain the principal source of many of the metal ores that are indispensable to capitalist industry.

Tables VIII and IX both reveal the extent to which the exports of the Third World countries are composed of raw materials, and the degree to which this composition has remained fairly constant, despite a slight decline in the exports of non-ferrous metals. The overwhelmingly dependent character of Third World economies' industrial sectors is

TABLE VII Third World production as a percentage of production in Europe[a] and N. America (1973)

Raw material	*Percentage*
Phosphates	49.7
Iron Ore	60.7
Bauxite	53.4
Copper	5[illegible]
Tin	96.8
Zinc	41.0
Lead	40.0
Chrome Ore	84.0
Petroleum	73.9

[a] Inc. E. European countries, except USSR
Source: UN Statistical Yearbook 1974

starkly revealed by the fact that approximately 50 per cent of their exports to the developed countries were still composed of raw materials and fuels in 1973.

TABLE VIII Trade between developing countries and rest of world, 1960–7

Importing regions	*% of Total Developing countries export 1959–61*		*1967–9*	
Developed Market Economies	94		93	
of which				
Foodstuffs	30		23	
Raw Materials	27)	52	19)	51
Fuels	25)		32)	
Manufactures	12		19	
Total	100		100	

Source: UN World Economic Survey, 1969–70, p. 137

Moreover, the raw material extraction and processing component of the industrial sector has been characterised by an extremely rapid increase in output. During the sixties the increase in output of the extractive industries has, as Table X indicates, been around 10 per cent

TABLE IX Industrial capitalist economies: major imports from underdeveloped countries, 1970–3

Commodity	1970 Value ($)	1970 % of Imports	1971 Value ($)	1971 % of Imports	1972 Value ($)	1972 % of Imports	1973 Value ($)	1973 % of Imports
Petroleum Products	14086	32.6	18343	38.1	20972	37.7	29903	37.3
Non-ferrous Metals	3359	7.8	2405	5.0	2459	4.4	3754	4.7
Metalliferous Ores	2985	6.9	2848	6.0	2917	5.2	4039	5.0
Wood Lumber	1275	3.0	1354	2.8	1530	2.8	3102	3.9
Total		50.3		51.9		50.1		50.9

Source: UN World Economic Survey, 1974, Table 43, p. 129.

per annum. Further investigation of particular raw material commodities reveals a relative decline in tin, chrome ore, zinc and lead ore, but this in no way negates the overall trend; increase in commodities such as bauxites, iron ore and petroleum have been spectacular, as Table X shows.

TABLE X Increase in output of particular raw material commodities 1948–1970 (in 000 tons)

Commodity	*1948*	*1960*	*1970*
Iron Ore	7,300	49,500	112,000
Bauxite	5,100	17,000	31,800
Petroleum	156,000	497,000	1,314,000

This level of extraction for export to meet the requirements of the industrial capitalist economies contrasts vividly with the degree to which Third World economies actually utilise the materials they extract for their own industrial production. Throughout the sixties Third World countries contained on average only 5 per cent of the world's heavy industry, and 10.5 per cent of its light industry.[11] Consider the disparity between production of extractive commodities and their utilisation in productive consumption, as revealed in Table XI.

TABLE XI Production and consumption of energy (in metric tons of coal or equivalent)

1964	*Production*					*Consumption*
	Total	*Coal*	*Petroleum*	*Gas*	*Electricity*	
U.S. & Canada	1,707	465	567	638	37	1,823
W. Europe	584	499	27	24	34	1,014
Japan	64	51	2	3	8	161
Australasia	38	36	0	0	0	56
Latin America	388	8	308	35	6	177
Middle East	568	6	557	5	1	53
Asia	128	79	39	6	4	136
Africa	97	49	45	1	1	71
% for Third World	22%	6%	51%	5%	11%	9%
For 1965	23%	7%	52%	5.5%	9%	8.8%

Source: Jalée (*TWWE*) p.42

Whilst this table only provides figures for 1964 and 1965, it nevertheless expresses the 'disparity' in a very succinct manner. Furthermore, since the mid-sixties overall trends have hardly changed, as can be seen from an examination of the U.N. Statistical Yearbook for 1974. Here, for example, in 1973 Africa produced 466.64 mt of energy, yet consumed only 142.95 mt; Latin America produced 436.19 mt and consumed 300 mt; the Middle East produced 1,426.00 mt whilst consuming a mere 107.42. The situation in Asia has deteriorated—whilst producing 250.51 mt, the continent consumed a relatively *lower* figure of 228.12 mt. (This, of course, excludes Japan, which produced a miniscule 39.73 mt, yet consumed a massive 390.20 mt.)

We can draw a number of conclusions from this data.

The crucial role that the raw material extractive sector plays in the economy is directed overwhelmingly to meeting the reproductive requirements of industrial capitalist economies, rather than its own productive consumption. It is a sector that has constantly experienced increases in rates of output during the post-war period, yet these have done little in general terms to boost the economic development of industries geared towards the domestic market, as revealed by the consumption of energy data. Its higher levels of productivity, advanced capital-intensitivity, considerably higher rates of accumulation and mass of profit, stand in stark contrast to the manufacturing sectors of Third World economies. There is, again, a marked sectoral unevenness here, between the two major components—the extractive and the manufacturing.

Furthermore, as we shall see from our examination of the manufacturing sector, this unevenness is a result of the way in which economic penetration by the industrial capitalist economies has structured the industrial sector of Third World countries into a form that restricts its possibilities for directing production towards the requirements of the indigenous economy; capitalist penetration has directed the latter away from its fundamental objective—of ensuring the enlarged reproduction of an economic system that can establish a basis for interrelating the extractive, manufacturing, and agricultural sectors into a unified entity geared to the domestic market.

The manufacturing component of the industrial sector

The disparity between production of energy and its internal domestic consumption is—as indicated above—considerable. Whilst, for example, the Third World economies produce all the world's natural rubber, they only consume approximately 18 per cent of this commodity; similarly, whilst they extract around 60 per cent of the phosphates that are consumed by the industrial capitalist economies, they only consume 6 per cent of the world's production. This low level of consumption of raw material products indicates the limited development of manufacturing industry in the Third World. Whilst raw materials and basic foodstuffs constituted approximately 80 per cent of exports, manufactured goods formed only 10 per cent of the total throughout most of the 1960s. Conversely, 60 per cent of the imports of Third World economies were manufactured goods during the same period. Furthermore, the growth that took place in the manufacturing sector during the fifties and sixties was lower than that of the extractive sector: compare, for example, the annual average rate of growth of 9 per cent for petroleum production in this period, or the average growth between 6 and 8 per cent for mineral extraction with the per capita growth rates in Table XII.

In addition to levels of output, the manufacturing sector also has a much lower productivity relative to the extractive sector; it goes without saying, of course, that it is also lower than the manufacturing sector of the industrial capitalist economies.[12]

This unevenness between the two compoments of the industrial sector, measured by growth, productivity and low levels of manufacturing export, and by the subsequent dependency of the Third World economies on manufactured imports, is, as we shall see below, a result of the way in which capitalist penetration has restricted and confined industrial development to certain sectors, to the detriment of others.

TABLE XII Annual rate of growth of manufacturing industry

Total non-communist less developed-countries	*Output manufacturing industries*	*Population*	*Output manufacturing industry per capita*
1938–1950	3.8%	1.7%	2.1%
1950–1960	6.9%	2.3%	4.5%
1960–1970	6.3%	2.6%	3.6%

Total non-communist developed countries	*Output manufacturing industries*	*Population*	*Output manufacturing industry per capita*
1938–1950	4.5%	0.3%	3.8%
1950–1960	5.0%	1.2%	3.8%
1960–1970	5.6%	1.1%	4.4%

Source: P. Bairoch, op. cit., p. 67.

Within this framework, there have, however, been important changes in the manufacturing sub-sector during the late sixties and early seventies and, before going on to examine the way in which industrial capitalist penetration has promoted restricted development in this area, we should briefly examine these changes, since they have led a number of authors, such as Warren,[13] Emmanuel,[14] and others to conclude that Third World countries—*in general*—are currently embarking on an autonomous industrialisation which heralds the beginning of the end of industrial capitalist domination.

What, then, are these changes? Take Table XIII.

There has undoubtedly been a major average increase in the output of manufacturing industry in Third World economies during the period from 1967. To make such a statement, however, is to say little. What we need to investigate are the particular sub-sectors in which the increase in output has occurred, the specificity of those economies which have experienced particularly high rates of manufacturing growth, and the extent to which this growth remains dependent upon penetration by the industrial capitalist states. Only by investigating these issues will we be able to go beyond the level of statement to analyse the extent to which the development of manufacturing industry in Third World economies remains confined and restricted to certain sub-sectors, to the exclusion of others.

Broadly speaking, the increase in manufactured output has occurred in three forms: import-substitution, the setting up of linkage industries

TABLE XIII Index numbers of Industrial Production (1963 = 100)

	1960	*1961*	*1962*	*1964*	*1965*	*1966*	*1967*	*1968*	*1969*	*1970*	*1971*	*1972*	*Annual Growth Rate 1971–3*
Developed market economies	85	88	94	108	116	125	128	137	148	159	152	163	6.2
Developing economies	86	92	96	110	118	123	129	138	150	158	170	185	6.7
of which:													
Food, beverage & tobacco	91	94	96	106	112	119	123	128	137	143	150	160	4.8
Textiles	93	97	97	108	111	111	116	126	131	139	148	162	4.0
Wearing apparel	86	92	93	106	116	123	129	136	134	135	144	151	6.8
Wood products	83	85	92	107	129	126	132	154	155	149	162	169	5.5
Paper and printing	79	88	94	105	112	118	121	135	146	157	172	189	8.5
Paper, paper products	73	84	90	110	120	128	132	145	155	168	181	195	7.4
Chemical petroleum products	84	91	96	110	117	125	133	148	162	176	181	210	7.5
Petroleum & coal products	85	89	97	107	109	117	127	137	146	158	162	169	5.4
Non-metallic mineral products	84	89	96	109	117	123	132	143	157	171	189	202	7.9
Basic metal industries	79	85	91	112	118	130	131	146	159	164	175	194	8.2
Metal products, machinery	76	89	97	119	133	137	141	159	181	195	214	143	10.0
Electrical machinery	77	92	96	117	133	151	160	188	210	233	257	287	10.9
Transport equipment	77	90	102	120	129	132	134	152	182	199	211	242	10.6
Rubber and plastic products	89	99	102	111	122	127	131	146	158	167	184	192	6.7

Source: UN *The Growth of Industry*, Vol. 1, Editions 1972, 1973, Pt. 2, Table I

tied to raw material extraction and processing industries, and assembly operations under the (overt or covert) control of multi-national corporations. Although we deal with these issues later in the text, it is important here to briefly state how these various forms generate an increasing dependence on capitalist penetration to the detriment of production for domestic requirements, and how they produce an increasing unevenness within the manufacturing sector.

Import-substitution entails the setting up of industrial units of production producing consumer goods for the domestic market. Concomitantly, high tariffs are placed on the importing of these goods into the country. Since the whole process is dependent on a replication of the technological forms by which these commodities are produced in the industrial capitalist economies, the new units are dependent upon the latter for the importing of machinery, parts, and so on—as in the 'production' of cars from imported parts, the use of imported machinery to produce selected garments, etc. Indeed, the industrial capitalist economy will usually only agree to the erection of the tariff if it is accompanied by a continuation of machine imports. Consequently, the possibility of establishing backward linkages from the import-substituting industry is blocked at the outset, and we have a relatively more advanced industrial enclave unrelated to other economic sectors, with whom it cannot be tied in any sustained indigenous economic growth. Furthermore, if the possibility arises for protecting other related sectors, development of production then requires an increased rate of investment, which can be generated either by increasing the amount or price of exports from the primary sector, or by increasing taxation internally. Whilst the former policy runs the risk of antagonising the very suppliers of means of production on whom the process of import substitution must be based in the short term, the latter results in a cut-back on the domestic consumption that is so vital for the continuing existence of import-substitution. Consequently, it is hardly surprising that, whilst the introduction of import-substitution has led to considerable increase in manufacturing output in the short-term (for example, in Argentina and Chile, in the mid-fifties, Brazil in the sixties) it has ultimately come up against the limits of the domestic market and the difficulties of internal backward linkages.

The development of manufacturing industries tied to the extractive sector is similarly a means for reinforcing unevenness and dependence. The setting up of such units of production, processing, for example, by-products from raw material extraction,[15] entails the creation of industries directly linked to and dependent on the foreign-dominated extractive enclave for provision of equipment, technical knowledge, etc., whilst the only possible forward linkages are a diversification of dependent processing, or distributive agencies (which often market the product in another country where a higher rate of profit prevails). At the same time, the extractive sector benefits from the provision of local services developed by the state or the local entrepreneur for the burgeoning national industries.[16] Consequently, whilst the increase in output within this area can be considerable,[17] it contributes little to a process of self-sustained economic growth geared to the requirements of the domestic market, and based on an indigenously balanced inter-sectoral development of capital and consumer goods industries.

Since the late sixties, large multi-national corporations have been vigorously instigating a qualitative change in the international division of labour within the framework of the world capitalist economy. Under the control of a single centre of economic ownership, production units are set up in different countries, each based on a particular labour process not integrated with other processes in the domestic economy, but rather within a division of labour that *inter-relates labour processes in different economic formations*. This inter-relation can take varying forms. A particular subsidiary (which can be directly owned by the multi-national, jointly owned by it, or even domestically owned) in one country can be responsible for producing one or several components of a commodity; alternatively, any one subsidiary can specialise in all stages of the production of a commodity, which is then exchanged for another commodity produced by a subsidiary in another country (e.g. production and exchange of car components).[18] Similarly, a large multi-national corporation producing a variety of commodities in any one sector (e.g. electrical engineering) may well choose to site its less capital-intensive labour processes (e.g. the assemblage of pocket calculators or T. V. sets, from parts produced elsewhere) in areas where labour power is relatively more abundant and cheaper. Hence the predominance of assembly operations in countries such as Hong Kong, Taiwan, South Korea, and so on.

This process of international inter-relation of labour processes in the production of a commodity under the effective control of an economic ownership that is trans-national has rapidly increased since the mid-sixties. Indeed, it accounts for much of the boom in the manufacturing output of many Third World economies, as can be seen from the rapid increase in the output of 'electrical machinery' and 'metal products' (Table XIII). However, as with the other elements of the boom outlined above, this process reinforces the existing unevenness within the manufacturing sector, since its very basis is an industrial fragmentation. Each industrial enclave is restricted to participating in a part of production, without any access to or control over the production process as an entirety. Whatever form actually taken by the ownership of the unit of production participating in this process, control lies in the hands of those who supply the means of production and determine the distribution of the product. Any spread of such manufacturing can only occur within this overall framework, which produces a developing enclave which is quite unrelated to the requirements of the domestic economy.

If we now turn from the actual composition of this manufacturing to its *location*, the external dependency of the contemporary increase in manufacturing output in the Third World becomes even clearer.[19] Take for example, Table XIV.

TABLE XIV Leading developing countries in the export of manufactured goods, 1970 and 1972

	Value of Manufactured Exports ($ million)	
Country	*1970*	*1972*
Hong Kong	1,949	3,233
S. Korea	641	1,359
Pakistan	397	376
India	1,081	1,028
Mexico	391	647
Singapore	428	890
Egypt	205	256
Argentina	246	381
Brazil	363	736
Total: These countries	5,701	8,906
Total: All developing countries	8,905	15,623

Source: UNCTAD, documents TD/B/429/add. 2, p. 33, and TD/B/530/Add.I (Part I) Annex p. 27, cited in A. K. Bhattachraya, *Foreign Trade and International Development*, p. 39 (excluding data on Israel)

From this data, we can see that, in 1972, nine countries provided 57 per cent of the total exports of manufactured commodities from the Third World economies. Two of these, Hong Kong and S. Korea, have built their assemblage industries in the manufacturing sector by utilising massive amounts of foreign capital. India's increase in output is based largely on import-substitution in the textiles and clothing sector. Mexico's increase is, again, based on the introduction of assemblage and components production, established with considerable injections of U.S. aid. Argentina has relied heavily upon import-substitution and processing, whilst Brazil's import substitution assemblage and components industries have been achieved through a process of massive subsidisation funded by heavy U.S. debt-financing, which has enabled Brazil to sell its goods on the overseas market at approximately one-half of their domestic prices.[20] Furthermore, most of these countries (with the exception of Argentina) are characterised by cheap and abundant supplies of labour-power, minimal restrictions on profit repatriation and the capacity for foreign companies to operate relatively freely behind joint-owned or domestically-owned front companies.

Given such conditions, it is hardly surprising that, whatever form taken by the increase in manufacturing output, foreign investors are prepared to enter these economies with the object of a short-term rate of return which is reasonably high, relative to profit rates in other industrial capitalist economies—particularly when there are less advanced technological labour-intensive tasks to be carried out in any one process involved in the production of a particular commodity.

Examining the characteristics and geographical location of the recent increasing in manufacturing output, we can, therefore, conclude that this increase is confined to *three basic forms*—forms which are dependent on and often geared towards the reproduction of the industrial capitalist economies. These forms exist as dependent enclaves in the manufacturing sector, enclaves which are—at best—only marginally tied to the rest of the sector, and unrelated to the reproductive requirements of the domestic economy. The subsequent unevenness in this sector must thus be seen as a result of the confining or restricting of development to specific areas; to argue otherwise is to present a misleading picture, which ignores this specificity. As long as development is confined in this way, the possibility of promoting a balanced growth—whether it be capitalist or otherwise—between the two departments of industrial production, inter-related with a development of the agricultural sector, is blocked both by the policies of the industrial capitalist countries and by the resistance to such a growth by those classes whose economic dominance is based on the perpetuation of these policies in the industrial and agricultural sectors.[21]

Such then, is the nature of the manufacturing component of the industrial sector, a component whose development—as with the sector in which it exists—is unevenly and restrictively structured.

Uneven and restricted development in the industrial sector

Thus far, we have indicated briefly how capitalist penetration has produced an industrial sector whose configuration is characterised by a highly uneven development, which is confined to particular areas; these are either crucial for the reproductive requirements of industrial capitalist modes of production, or generate an increasing dependence on these modes for their own reproduction.

The Sociology of Development by its simplistic equation of patterns of industrial growth in the now industrialised capitalist states with the process of uneven and restricted development that is occurring in the extractive and manufacturing sectors of Third World economies is, therefore, committing a serious analytical error. The process of relatively balanced economic growth between industrial sectors that characterised capitalist industrialisation in Western Europe and the U.S. (as revealed, for example in Table XV for the U.S. during its most crucial industrialising period), is simply not in evidence in the 'industrialisation' of Third World economies; indeed, for the period, 1958–70, there is, overall, a marked correlation between high rates of growth in the industrial sector and low rates of growth in the manufacturing sector.[22]

TABLE XV Annual rates of growth of output for U.S.A.

Period	*Manufacturing industry*	*Extractive industry*	*Ratio 2.1%*
1899–1914	5.5	6.0	109
1914–1924	4.3	4.3	100
1924–1929	4.8	3.7	77
1929–1951	3.4	2.4	71

Foreign trade

The sectoral configuration we have outlined above is, of course, reflected in the extremely unbalanced nature of trade between Third World and industrial capitalist economies.

Whilst the former are heavily dependent on the latter for their trade, the industrial capitalist economies are largely dependent on trade between themselves. This is evidenced in Table XVI, where, in 1973 74.2 per cent of Third World exports went to the industrial capitalist economies and only 19.9 per cent to other Third World countries, whereas only 18.1 per cent of industrial capitalist exports went to the latter, whilst internal trade between the capitalist economies amounted to 76.4 per cent. In 1966, the dependence of these groupings of economies on their trade with the other was 72.8 per cent for the underdeveloped and 20.2 per cent for the developed, whereas in 1973 it was 74.2 and 18.1 per cent respectively.

Consequently, it seems that in trading terms, Third World economies are becoming more dependent on the industrial capitalist economies, which, in turn, are less dependent on the underdeveloped. The contemporary situation is aptly summarised by Paul Bairoch:

> We are confronted with a pattern which, in global terms, places the underdeveloped countries in an unfavourable situation, for although the western market is of prime, even vital, importance to them, the place of the underdeveloped countries in the trade of the western world can by contrast be considered if not exactly marginal, then certainly far from being of major importance.[23]

That this 'unfavourable situation' in external trade is rooted in the uneven and restricted economic development particular to Third World economies can be seen from Table XVII, showing foreign trade by commodity groupings.

From this it can be seen that whilst manufactured goods formed 23 per cent of exports, in 1970 they comprised 69 per cent of imports. In the same year, manufactured goods formed 76 per cent of exports from the industrial capitalist economies. Furthermore, since, as we noted above, the 23 per cent of exports came largely from a small group of countries, this average conceals a highly selective growth. Meanwhile, exports of fuels from Third World countries continued to comprise a massive 33 per cent, whilst all raw material export from both the agricultural and extractive industrial sectors totalled 75.8 per cent. From this, we can draw the following conclusion.

In trading terms, the dependence of Third World upon industrial capitalist economies appears to be increasing. Contrary to those arguments which deduce a decline in the dependence of the economic structure from the phenomenon of price rises in the extractive sector,[24] the pattern of trade reveals an increasing dependence on production to meet the reproductive requirements of the industrial capitalist

TABLE XVI World exports in 1966, 1970, 1973 by groups of countries and by origin and destination

Country		Total exports	Developed capitalist countries		Europe, USSR, Vietnam, Korea China, Mongolia		Underdeveloped countries	
			Total	%	Total	%	Total	%
U.S. & Canada	1973	95430	69830	73.2	3160	3.3	21760	22.8
W. Europe	1973	258610	209760	81.1	13780	5.3	33140	12.8
Australia, New Zealand S. Africa	1973	12850	9850	76.7	513	4.0	2393	18.6
Japan	1973	36930	19140	51.8	1960	5.3	15830	42.8
Total developed countries	1966	141710	105250	74.3	6110	4.3	28650	20.2
	1970	224840	172460	77.4	9020	4.0	41910	18.6
	1973	405320	309710	76.4	19430	4.8	73380	18.1
E. Europe	1966	23500	5830	24.8	13860	58.9	3690	15.7
USSR, etc.	1970	33360	8040	24.1	19920	59.7	5180	15.5
	1973	57720	16340	28.3	32360	56.0	8730	15.1
L. America	1973	29080	22030	75.7	1590	5.5	5050	17.3
Africa	1973	20280	16430	81.0	1510	7.4	2120	10.4
Middle East	1973	27000	20180	74.7	670	2.5	5690	21.1
S. & S.E. Asia	1973	31160	21410	68.7	1260	4.0	8170	26.2
Total underdeveloped	1966	38390	27940	72.8	2390	6.2	7680	20.0
	1970	55010	40660	73.9	3130	5.7	10710	19.5
	1973	108840	80810	74.2	5040	4.6	21190	19.4
World	1966	203600	139020	68.2	22370	10.0	40010	19.6
World	1970	313200	221160	70.6	32070	10.2	57810	18.4
World	1973	571520	406860	71.2	56840	9.9	103300	18.1

Source: P. Bairoch, p. 99

TABLE XVII Foreign trade by commodity groupings of non-Communist less-developed countries

	Imports			Exports		
	1955	*1960*	*1970*	*1955*	*1960*	*1970*
Total (U.S. $ Billion)	23.1	29.1	57.7	23.7	27.4	54.3
Distribution of products as percentage of total						
1 Food products	15.4	16.2	12.9	32.5	29.6	24.3
2 Primary products	8.1	7.4	6.2	29.4	27.9	18.2
3 Fuels	11.8	10.0	8.0	25.2	27.9	33.3
4 Chemicals	7.0	7.5	8.9	1.0	1.1	1.5
5 Machinery	23.3	27.6	33.4	0.5	0.7	2.4
6 Other manufactures	29.0	27.8	26.4	11.4	12.4	19.5
1–3 Raw materials	35.3	33.6	27.1	87.1	85.4	75.8
4–6 Processed goods	59.3	62.9	68.8	12.9	14.2	23.4

Source: P. Bairoch, p. 98, Table 31

economies in the agricultural and extractive sectors, to the detriment of the remainder of the economic structure. Within the latter, any overall increase in the production of commodities for export is based on an importing of capital goods for manufacturing or processing industries in sectors whose reproduction—as we have seen—remains structured by the requirements of capitalist penetration. *The increase in dependence of external trade on the industrial capitalist economies* seems, therefore, to be *contemporarily reinforcing* the pattern of uneven and restricted development that characterises the economic structure of Third World economies. The marked contrast between the trading patterns of the latter and those of the industrial capitalist economies starkly reveals the degree of sectoral unevenness dramatically specific to Third World economies.

Commodity exchange

In addition to this increasing trading dependence, the structuring of the Third World economy in terms of the reproductive requirements of the industrial capitalist economy also produces, at a more basic level, *a particular form of commodity exchange* between these economies.

This particularity is, again, a phenomenon that is ignored by the Sociology of Development, with its implicit assumption that this exchange is equivalent to that between the industrial capitalist states, where, supposedly, each country exports the product it can most adequately produce, and benefits from an international division of labour which ensures that it can import the commodities other countries specialise in exporting. Far from this exchange being characterised by a 'mutual' or 'comparative' advantage, it is, rather, based upon a *fundamental 'disadvantage' for Third World economies*, a disadvantage that results from the conditions under which production takes place in those sectors whose reproduction is dependent on capitalist penetration, or whose products are geared to industrial capitalist reproductive requirements.

A number of economists have argued that an inequality in the trade between the Third World and industrial capitalist economies can be directly revealed by an examination of the terms of trade and the balance of trade between them. Basing his conclusions on trade up until the mid-sixties, Jalée, for example, considered that the increase in the prices of manufactured commodities relative to the prices of primary products, and the increasing dependence of the Third World on industrial capitalist markets has led to a trend towards a deterioration in the terms of trade of the former *vis-à-vis* the latter, and, consequently, towards a worsening in the balance of trade of Third World economies. Citing figures such as a 4 per cent decline in the unit value of exports for underdeveloped countries, compared with a 19 per cent rise in the unit value of exports from industrial capitalist economies for the period 1950–64, Jalée concludes that changes in the pattern of trade have

> . . . served to strengthen objectively the imperialist character of the relations which subordinate the Third World to the advanced capitalist countries.[25]

If we examine recent data, however, as, for example, in Table XVIII, the contrast with this conclusion is quite striking. Here we see a massive improvement in the terms of trade for Third World economies and a concomitant deterioration in the terms for industrial capitalist economies. Similarly, if we examine data on the balance of trade (Table XIX), a trend towards an 'increasing deficit' in the general categories chosen by Jalée (Latin America, Africa, Far East, Western Asia) is not in evidence.

Jalée's conclusions, in fact, only appear to be valid for a period limited largely to 1952–62, which was then followed by a remarkable

TABLE XVIII World trade of market economies index by unit price ($ U.S) (1970 = 100)

	Exports							*Imports*							*Terms of trade*						
	1960	*1965*	*1971*	*1972*	*1973*	*1974*	*1975*	*1960*	*1965*	*1971*	*1972*	*1973*	*1974*	*1975*	*1960*	*1965*	*1971*	*1972*	*1973*	*1974*	*1975*
World	88	91	105	114	141	199	216	90	93	105	114	140	198	216							
Developed market economies	96	90	105	114	138	172	192	90	92	106	114	140	197	215	96	98	99	100	99	87	89
Developing market economies	94	93	106	115	155	310	325	91	94	105	114	140	199	227	103	99	107	107	111	156	147

Source: UN Monthly Bulletin of Statistics, April 1975

Table XIX World trade by region ($ U.S.)

Region	Imports (c. i. f.)						Exports (f. o. b.)					
	1958	*1966*	*1970*	*1971*	*1972*	*1973*	*1958*	*1966*	*1970*	*1971*	*1972*	*1973*
Latin America	10490	13110	18940	21200	23780	29980	9600	13380	17190	17720	20540	28900
Africa	6020	8230	11020	12580	13560	17710	4650	8200	12600	12920	15170	20570
W. Asia	2670	4860	6360	7480	9420	12920	4070	7260	10590	15260	19030	27020
Far East	8220	13960	19010	21030	23490	33960	6420	10020	14440	16120	20200	31300
Oceana	190	510	900	1000	1040	1200	140	260	480	520	640	970
Total: Underdeveloped	27600	40700	56200	6330	71300	95800	24900	39100	55300	62500	75600	108800

Source: UN Statistical Yearbook, 1974, Table 148

improvement—(at the level of 'average')[26] in both the terms and balance of trade. Thus, the idea of basing one's analysis of commodity exchange between Third World and industrial capitalist economies on long-term price fluctuations and deducing from this an 'objective strengthening' of the latter's dominance is obviously inadequate.

Beyond this, the method in itself is also limited; no matter whether or not the prices of goods from the extractive or industrial sector are increasing or decreasing per unit relative to imported manufactured goods, there remain major differences in the process of production *vis-à-vis* the industrial capitalist economies—differences which, *regardless of price fluctuations in the short term*, have, as a result of capitalist penetration, created a permanent 'disadvantage' for the Third World economies, no matter in what price form the end-result of the process of production is expressed.

As is well known, the major differentiating characteristic between the production processes in industrial capitalist and Third World economies lies in the differing levels of the development of the productive forces relative to each economic structure. Taken as a whole, the organic composition of capital in industrial capitalist economies is, of course, higher than that in Third World economies. The lower organic composition of capital in the latter makes possible an unequal exchange of commodities, since the production costs per unit will, on average, be much higher in the Third World economy. Consequently, we have a situation in which, in general, commodities containing more labour-time produced in the Third World economy are exchanged for commodities containing less labour-time produced in the industrial capitalist economy. There is, therefore, in value-terms, an exchange of non-equivalents. The higher cost of production in the Third World economy enables commodities produced in the industrial capitalist economy to be sold above their value, yet still below the value of domestically produced commodities. Thus, most commodities produced in the Third World economy can be constantly undercut by commodities produced at a lower cost in the industrial capitalist economy. The fact that the organic composition of capital is lower also means that the value of labour power is lower, since this is determined by the level of productivity in the wage-goods sector. The resultant supply of cheap labour makes it more profitable for foreign companies to locate production that requires a high level of labour-intensity in Third World rather than in industrial capitalist economies; the amount of labour-time required to assemble components in the former may be greater than in the latter, yet the price at which the product can be sold is lower than would be possible in the industrial capitalist economy.

Consequently, we can see that taking production as a whole, there is, as a result of the relatively differing levels of development of the productive forces, a massive transference of value from the Third World to the industrial capitalist economies as a result of an unequal exchange *rooted in the production process itself.*

This conclusion that there is a constant basis for a transference of value in favour of industrial capitalist economies should not be confused with Emmanuel's, notion of 'unequal exchange'.[27] According to his analysis, prices are formed internationally in such a way that the factors of production (land, labour, capital) are 'unequally rewarded'—particularly labour, since its price is much lower in the Third World economy. He argues that, due to *inequalities in wage levels*, a commodity produced in a certain number of hours in an 'underdeveloped' country can be purchased by a 'developed' country which exchanges it for a commodity produced in a smaller number of labour-hours; this occurs because, in the world market, prices oscillate around their monetary cost, and not their labour cost. For Emmanuel, therefore, the value of labour is the crucial determinant, the 'independent variable' that ensures unequal exchange. On the contrary, we are arguing that this determinant is not 'independent' but is formed by the productivity of labour in the wage sector, which is itself a result of the general level of organic composition of capital, and that the latter is the crucial determinant of the 'inequality'. Posing the question of the relatively lower organic composition of capital necessarily raises the problem of the productive forces whose development requires a particular level of composition; it entails enquiring how the dominance of existing relations of production in Third World economies perpetuates a restricted development of their productive forces. This, however, takes us into areas that will be dealt with later in the text. It is sufficient, therefore, simply to note here the conditions of production that have created a constant 'disadvantage' for Third World economies.

The uneven and restricted development characterising Third World economies is, of course, reinforced by the monetary policies—the so-called financial 'aid'—of the industrial capitalist economies. The general conditions under which an industrial capitalist state uses its own tax appropriations to guarantee markets in these economies has been fully documented elsewhere,[28] and there is no need to reiterate these arguments here. The overall conclusions of these analyses, that a reliance on aid tends to undermine indigenous industries, strengthen foreign-controlled sectors, and favour agricultural export against domestic production, clearly reveal how the imposition of monetary policies promotes an increasing sectoral unevenness.

CONCLUSION

Our analysis has shown that in each area of the economic structure examined, there exists a pattern of development characterised by a highly uneven sectoral configuration, in which a dominant yet externally dependent sector restricts the development of other sectors. This has been the case in the three major areas we have selected—the agricultural, extractive and manufacturing industrial. Trade relations between industrial capitalist and Third World economies constantly reinforce the reproduction of this restricted and uneven development, and its effects are manifest in a process of commodity exchange which enables a major transference of value from the Third World to the industrial capitalist economies.

The discourse of the Sociology of Development is, as we have seen, founded on a denial of the specificity of this economic structure. Thus, in addition to its theoretical inadequacies in explaining the structure, reproduction and process of change in the social formations of the Third World, we must now add two major failings originating directly from its denial of the specificity of its object of analysis.

First, by equating the process of—more or less—balanced inter-sectoral growth that characterised previous forms of capitalist industrialisation with the form of industrialisation occuring in Third World economies, the Sociology of Development is guilty of a serious error. Analyses carried out within this framework can, of course, always 'discover' indices or construct comparisons, equating present with past average 'levels of growth', 'levels of capital-intensity', 'output of manufacturing industry' and so on. Yet behind these empiricist statistical averages—which are no more than self-fulfilling prophecies designed to validate its basic assumption at the most general level—lies the very reality that denies such a simplistic transposition.

Secondly—and far more seriously, since the problem can no longer be confined to the academic level—the necessary corollary of policy recommendations based upon the theoretical conclusions of the Sociology of Development must be, as a result of its basic tenets to promote an extension of the very form of restricted and uneven economic development which, as we have seen, constantly blocks the emergence of a system of production directed towards the requirements of the domestic market. In so doing, such recommendations further ensure the structuring of the Third World economy to the reproductive requirements of the external capitalist mode of production, to the detriment of the domestic economy and the economic needs of the indigenous populations, whose overall dependence on the industrialised capitalist states is thereby further reinforced.

3 The Sociology of Under-development: the Theories of Baran and Frank

Turning from the conventions of the Sociology of Development, we can now examine its theoretical alternative—the discourse which has come to be known as the Sociology of Underdevelopment.

As distinct from the Sociology of Development, the Underdevelopment theorists—Baran,[1] Sweezy,[2] and Frank[3] (hereafter referred to as BSF)—recognise both the restrictive effects of capitalist penetration of Third World economies, and the fact that these effects vary from one period of capitalist development to another.

Yet, despite this recognition, it seems to me that their discourse contains serious explanatory limitations, which ultimately render it theoretically untenable. Since research is increasingly relying on conclusions reached by these authors, it is essential that—as with the Sociology of Development—we clearly demarcate the basic elements of underdevelopment theory, assessing its adequacy in analysing its theoretical object.

THE DISCOURSE OF UNDERDEVELOPMENT

The concept of economic surplus

The concept of economic surplus and its generation and absorption form the explanatory essence of the BSF discourse. Despite this, surprisingly little attention is given to its definition. In *The Political Economy of Growth*, Baran defines it as, 'the difference between society's *actual* current output and its *actual* current consumption.'[4] This definition, applicable to all modes of production (since they all generate a surplus), is retained in an abbreviated form in Baran and Sweezy's *Monopoly Capital* where surplus is defined as, 'the difference between what a society produces and the costs of producing it.'[5]

The first point to be made with regard to this definition is the 'obvious'—that what society produces (i.e. the use-values that are produced for productive and personal consumption) varies from one mode of production to another. Similarly, the costs themselves also vary. For example, whilst the cost-price of a commodity produced in a capitalist production process embodies the cost of the elements consumed in production—machinery, raw-materials, labour, etc.—the cost-price of a commodity produced by a tenant farmer under feudal relations of production would be composed of ground rent, individual means of production, and his individual assessments of the monetary returns sufficient to reproduce the necessary conditions for commodity production, and so on. What a society produces, and the cost of producing this—which can only be subsumed under the question of *how* it is produced—is the result of the structure within which production itself takes place, which as we shall see in detail later in the text differs from one mode of production to another. What Baran and Sweezy regard, then, as the determinants of the surplus are themselves determined by a specific combination of relations of production and productive forces that exists as a particular mode of production. To take an example: in the capitalist mode of production, the preconditions for the existence of a capitalist cost-price are a separation of direct producers from their means of production, a division of labour which results in separate units of production producing commodities as exchange-values for the market, and so on. Similarly, the production of use-values in the capitalist mode is determined by such factors as the rate of exploitation and the level attained by the development of the productive forces, which requires a particular composition of means of production in each branch of production. These specific determinants of the production of use values are, of course, themselves determined by the way in which the dominant capitalist relations of production distribute the social labour-force in the production of commodities.

It would appear, therefore, that the difference between the 'costs' of production and 'what' is produced is an *effect of the structure* of production; the 'determinants' of the surplus are themselves subject to a structural determination that is absent in the BSF discourse.

What, then, does determine that a surplus is produced in a mode of production? Having defined the 'economic surplus', Baran and Sweezy then go on to assert that, in all societies, a surplus *is* produced. The crucial question should then become the mode of extraction of this surplus-labour, the dominant relations under which this occurs. As we shall see, in any mode of production, the extraction of surplus-labour produced in particular labour processes is always governed by a specific set of relations of production. If we are to state, as Baran and Sweezy do, that the existence of a surplus and its 'utilisation' form the basis for an analysis of different social formations, then there must be a basis for distinguishing between the different forms of surplus-extraction characteristic of different modes of production. Yet, this is quite absent from their work since these different forms are totally subsumed under one—the 'existence' of a surplus. Whilst they argue that different forms of society utilise the surplus they produce in different ways, and then describe these, they provide no adequate basis for analysing this 'utilisation'. The structures of modes of production and their processes of reproduction, which are the determinants of differing forms of 'utilisation' they outline, remain unanalysed.

Because Baran and Sweezy ignore these determinants at the outset of their analysis, their investigation of underdevelopment necessarily generates a series of problems which restrict its validity for theorising both the development of Third World societies and the process of industrial capitalist penetration of these societies. Having reached this point, we can now turn to these problems.

Before doing this, however, it is necessary to outline a further basic notion which co-exists with the concepts, surplus and surplus utilisation.

Surplus absorption

In their analysis of monopoly capitalism and the necessity for capitalist penetration of non-capitalist economies, having defined the economic surplus, Baran and Sweezy make a second assumption that

> . . . the 'normal' state of a monopoly capitalist economy is stagnation, because, twist and turn as one will there is no way to avoid the conclusion that monopoly capitalism is a self-contradictory system. It tends to generate ever more surplus, yet it fails to provide the consumption and investment outlets required for the absorption of a rising surplus and hence for the smooth working of the system.[6]

Here we have a further axiom of the BSF discourse—that enlarged reproduction is impossible within a capitalist system, due to a lack of effective demand. The impossibility for the system to absorb an ever-rising amount of surplus is deduced from this latter assumption. The question is then posed as to what are the stimuli, external to this system, that can alleviate this stagnation by 'stimulating' demand. The system's dominance of underdeveloped economies is then analysed as one such 'stimulus'. The argument for the existence of this dominance is, however, based on an axiom whose validity is questionable. Can we, in fact, conclude that, within the capitalist economy, enlarged reproduction is constantly limited by the level of consumption? Can we simply analyse industrial capitalist dominance of Third World economies as an attempt to resolve problems produced by this?

In order to examine this, we need to analyse the status of consumption in the capitalist mode of production. As we noted in Chapter 1,[7] in many economic theories, consumption is defined in terms of the economic needs of a pre-given consuming human subject, and the means by which these needs can be 'satisfied'.

Contrary to this anthropological conception, however, we would argue that under capitalist production 'needs' cannot be defined in terms of 'human nature' (be it historical or a-historical), but must be analysed in relation to the effectivity of the structure of production. The process of production determines specific needs, since it produces both the use-values that are consumed and their mode of consumption (both individual and productive).

> Hence production produces consumption 1) by providing the material of consumption; 2) by determining the mode of consumption; 3) by creating in the consumer a need for the objects which it first presents as products. It therefore produces the object of consumption, the mode of consumption and the urge to consume. Similarly, consumption produces the predisposition of the producer by positing him as a purposive requirement.[8]

This, however, is the most general statement possible, and, in order to understand the nature of the relationship between production and consumption, it must be situated within the *development of the capitalist mode* of production. An increase in means of production produces a growth in the amount of means of consumption, and this increase also creates a new demand for the means of consumption. As a result, a particular level of production of means of production corresponds to that of the production of means of consumption. In the early stages of capitalist development, for example, the thesis that enlarged reproduction cannot take place due to insufficient demand, is clearly inapplicable—since capitalist production can create a home market for itself through the development of the social division of labour. The separation of direct producers from their means of production is the basis for this development, since it converts these means of production into capital which is then set in motion by the separated producers (now selling their labour-power to a capitalist) to produce commodities, which can be used either in productive or personal consumption. The market for both types of consumption expands: personal—through the existence of a labour force that can now only subsist by selling its labour-power and purchasing commodities produced by capitalist production; productive—by the increase in the demand for means of production in the separate units of production creating a market for means of production. At this stage productive consumption increases more rapidly than personal consumption, since a surplus of means of production must be available to increase overall production. Thus, at this stage, the growth of the home market occurs largely through productive and not personal consumption.

With the progressive subsumption of the productive forces under capitalist relations of production, analyses of consumption must be situated within the dynamics of the capitalist mode of production. As we shall see in detail in Chapter 5, the development of the structure of the capitalist mode of production produces a tendency for the rate of profit to fall; this tendency is an effect of the increasing organic composition of capital required for the development of labour productivity. It is, however, a tendency whose influence is checked by a series of counteracting forces—for example, an increase in working hours, a cheapening of raw materials and machinery, a reduction in wages relative to the value of output, and an increase in foreign trade—to name but a few. The relation between production and consumption at any given moment in the development of capitalist production must be analysed in relation to the operation of this tendency and its counteracting effects.

Let us take, for example, the counteracting effect of foreign trade. Through the importing of raw materials and agricultural commodities, foreign trade cheapens the elements of constant capital and also the subsistence goods for which variable capital is exchanged. Thus it can enable the rate of surplus-value to be raised, and the value of constant capital to be lowered, thereby raising the rate of profit in the short term. The rate of profit can also be raised by investing capital in countries with inferior production facilities. The resultant increase in accumulation and output from participation in foreign trade can then also produce an increase in consumption—wages can rise as a result of the increase in output for an overseas market, whilst the prices of commodities that are essential for the maintenance of labour-power can remain constant or fall. Thus, we basically have an increase in production and consumption, but the increase in the former is relatively much greater than in the latter. The level of consumption is transformed by the operation of a counteracting effect. This would be much the same if we had chosen one other, or a combination of effects; in each case, the level of consumption is a function of the operation of these effects at a particular moment in the development of capitalist production.

The underconsumptionist now analyses this situation, and reaches the conclusion that, since output is greater than demand, there must exist a 'permanent' state of underconsumption in the capitalist mode itself. As opposed to the structural determinants of the increase in output and consumption, the theory can see only the effects of this determination. The notion that there exists a contradiction or antagonism between the levels of production and consumption appears to be deduced from *a description of effects produced as a result of the dynamics of the capitalist mode of production*. The presumed contradiction is a general effect which, whilst not prevalent during the early stages of capitalist production, is produced as a result of the progressive limitation of the productive forces by capitalist relations of production, as represented in the movements of the tendency of that mode and its counter-effects. Baran and Sweezy take this general effect produced by the structural development of the capitalist mode of production as the basis for their analysis, without examining how it is itself determined and transformed by the reproduction of the structure. Consequently, their contradiction which supposedly establishes the 'necessity' for capitalist penetration of non-capitalist economies is a very limited foundation, since the phenomenon they are attempting to explain is itself determined by the development of a structure which also determines what they view as the 'cause' of the phenomenon. The inadequacy of this proof of the necessity for capitalist penetration is thus a major problem for their theory. Yet, more important for us are the *results of this proof* for the analysis of Third World societies. These can be briefly specified here, to be developed further in the analysis that follows.

Firstly, from what we have said above, it follows that the determinants of different forms of capitalist penetration of Third World societies since the pre-colonial period tend to be conflated into one major cause—the tendency to underconsume. The economic determinants of, for example, sixteenth-century Spanish colonial expansion in Latin America have to ultimately be the same as those of the European scramble for colonies in Africa at the end of the nineteenth century. The specific characteristics of a particular form of economic dominance of non-capitalist societies thus disappear behind general observations on the 'necessary expansion' of capital.

Secondly, as we shall see, these different forms of penetration had very different effects on the non-capitalist societies that were to become the underdeveloped formations of the Third World. For the Underdevelopment theorists, however, these different effects tend to remain unanalysed, since their specificity is less important than their overall cause. Thus, for example, Frank can conclude:

> One and the same historical process of the expansion and development of capitalism throughout the world has simultaneously generated—and continues to generate—both economic development and structural underdevelopment.[9]

In the analyses of Underdevelopment theorists, both the causes and effects of the different forms of capitalist penetration tend to be relegated to a secondary position. This relegation is a necessary result of the conflation produced by their underconsumptionist axioms.

Having outlined the general limitations of the concepts of economic surplus and surplus utilisation and absorption, we can now examine the results of their use in the analysis of Third World societies.

The analysis of underdevelopment

The effect of the theoretical inadequacies inherent in these founding concepts are apparent in the initial analysis of underdevelopment put forward by Baran in 'Towards a Morphology of Backwardness'.

In this section of his text, *The Political Economy of Growth*, Baran initially poses a fundamental problem:

> It would be a fascinating task to follow up the evolution of the volume and employment of the economic surplus in the course of pre-capitalist development.[10]

The concepts he operates with, however, preclude an answer to this question, since they can only subsume very different non-capitalist structures under the general descriptive notions of surplus and its differential utilisation. As a consequence, the possibility of rigorously demarcating analytically the structures and reproduction of non-capitalist modes—a basic requirement for the theory of

Underdevelopment—is considerably reduced; furthermore, the problems subsequently generated by this lead Baran into a number of errors, which seriously limit the relevance of his theoretical work for analysing Third World societies. These errors can be briefly outlined.

Feudalism: the universal antecedent of capitalism

In 'Towards a Morphology of Backwardness', Baran states:

> The problem may best be approached by recalling the conditions from which capitalism evolved in both the now advanced and the now underdeveloped parts of the world. These were everywhere a mode of production and a social and political order that are conveniently summarised under the name of feudalism. Not that the structure of feudalism was everywhere the same.[11]

What, then, for Baran is the structure of this 'feudalism'? In the most general terms, we can define the feudal mode of production as one in which the surplus product is extracted from direct producers (who retain possession of their means of production) in the form of a ground-rent (be it rent-in-kind, labour-rent, or money-rent) levied upon them by those who have established a legal right to an appropriation of the surplus (feudal landowners, nobility, monarchy, etc.). There is thus a dislocation here between the conditions of production—the labour process—and the conditions of appropriation of the surplus-product. The direct labourer remains in possession of the means of production and the conditions necessary for his own (and his family's) subsistence, whilst the conditions of appropriation defined through the dominant relations of production require that his surplus-labour become the property of another—the feudal landlord. The surplus-labour is appropriated through the legally-based political intervention of the landowner within the processes of production. Thus, the surplus-labour is extracted from the direct producers 'by other than economic pressure'.[12] The disjunction between the two processes—of production and the extraction of surplus-labour—inherent in the structure of the feudal mode thus means that the intervention of the political is a precondition for the existence and reproduction of the feudal mode. Such—in the briefest possible definition—is the structure of the feudal mode. The analysis of its existence in a given historical period is minimally predicated on such a theoretical construction.

As opposed to such a theorisation, however, Baran simply asserts its presence, which enables him to reach the erroneous conclusion that, prior to industrial capitalist penetration, there was simply 'feudalism'—'erroneous', since, as we shall see later in the text, a variety of non-capitalist modes of production pre-existed capitalist penetration, modes whose specificity must be capable of analysis.

The emergence of capitalism: an unfounded abstraction

Having made this assertion, Baran adds another: that a 'confluence of processes'—an increase in agricultural output, a displacement of the peasantry and a consequent development of an industrial labour-force, the evolution of a merchant and artisan class, and the accumulation of capital—were emerging from feudalism to form the indispensable preconditions for the emergence of capitalism. In Baran's analysis, this was occurring everywhere, prior to the entry of industrial capitalism.

In order to establish his conclusions, Baran extracts several aspects from Marx's analysis of the European transition from feudalism to capitalism, assuming that these also pre-existed capitalist penetration of Third World formations. Yet, is such an extraction valid? The conclusion that 'the peoples who came into the orbit of Western capitalist expansion found themselves in the twilight of feudalism and capitalism',[13] cannot be established by simply restating Marx's indications on the European transition; rather, it requires an analysis of the structure and development of the various non-capitalist modes that pre-existed European entry, an analysis of the preconditions for the emergence of capitalist production in non-capitalist formations, and an analysis of the effects of capitalist penetration on non-capitalist modes of production. Only on such a basis can the transition to the state of underdevelopment be analysed. Baran lays great stress on the process of primitive accumulation of capital in non-capitalist modes: he argues that this enabled a transition to capitalism to occur prior to capitalist penetration. But why should this necessarily be the case? Vast accumulations of wealth occurred, for instance, in a number of S. E. Asian economies in the seventeenth and eighteenth centuries. These were also accompanied by the development of a merchant class, an increase in commodity circulation, export and the extension of the social division of labour. However, as we shall see, the mode of production existing in these societies blocked the development of capitalism, because no mechanisms existed for the separation of direct producers from their means of production—the basic requirement for the reproduction of capitalist production. It was not until the late nineteenth-century capitalist penetration of this mode that such a process of separation could take place. So how could these societies (which Baran would term 'feudal') be evolving towards capitalism? How could capitalist penetration, as Baran argues, block this development if the fundamental precondition for the existence of the latter was non-existent? These strictures on Baran's analysis similarly apply to most African societies prior to nineteenth-century colonisation, where 'feudalism', again, is not in evidence. The process of transition from non-capitalist to capitalist production in the history of contemporary 'underdeveloped' societies is not simply analogous to the transition as it occurred in Western Europe. Yet, it is this latter history which Baran extracts and then tries to 'locate' in all non-capitalist societies. As a result, the problems of the varieties of transition to capitalism in Third World societies, of the role that different forms of penetration played in this, and the effects that the establishment of capitalism as a dominant system has on these societies, are all problems that are neglected in Baran's analysis.

The blocking of capitalist development

This neglect produces a further problem, since Baran goes on to argue that capitalist penetration universally blocked the development of the capitalist mode of production that was already taking place in non-capitalist societies. We have indicated that it often seems more likely that the obverse is correct, namely that capitalist penetration, rather than blocking the development of capitalism, created the basis for this development. This point will be examined in considerable detail later, but here I want to focus on the corollary of this conclusion—namely, that the future economic development of underdeveloped societies has as its fundamental precondition a 'removal' of industrial capitalist dominance. This conclusion is really an inversion of the discourse and object of the Sociology of Development. It establishes the basis for the Sociology of Underdevelopment, which receives its only real foundation in Frank's analysis. We shall deal more fully with this later, but here it is necessary to examine Baran's actual arguments in greater detail.

In *The Political Economy of Growth*, Baran states:

> Regardless of their national peculiarities, the pre-capitalist orders in Western Europe and in Japan, in Russia and in Asia were reaching at different times and in different ways their common historical destiny.[14]

Consequently, if the non-capitalist formations of Asia, Africa and Latin America had been 'left to themselves', they

> might have found in the course of time a shorter and surely less tortuous road toward a better and richer society.[15]

It follows, therefore, that, if they *were* now 'left to themselves', this 'less tortuous road' could be travelled. Although Baran does not take Frank's position, that the development of the underdeveloped social formation has been and is almost totally determined by capitalist penetration alone, it is clear that he nevertheless lays the foundations for such conclusions, since he establishes the opposition, capitalist penetration—underdevelopment/absence of capitalist penetration—development.[16]

The conclusion then follows that capitalist penetration is largely 'responsible' for underdevelopment, for the blocking of development itself. Having reached this point, Baran can then incorporate this conclusion into his surplus-utilisation model. However, whereas in the analysis of monopoly capitalism and imperialism it was used as an underconsumptionist justification for the existence of capitalist penetration, it is now utilised to clarify a conclusion that has *already been reached.*

As in the analysis of capitalism and monopoly capitalism, Baran asserts that a surplus is produced, that its limits are determined in the same way as in other modes of production (output-costs, etc.), and that it is utilised in different ways. This he again does without enquiring into the structure that determines this surplus and its utilisation. If, for example, we ask how a system can develop in which the surplus that is produced is appropriated by a landowning class, a merchant class, a comprador class, foreign enterprises and the state in such a way that indigenous capitalist development is blocked; or what are the dominant relations of production that structure this process of surplus extraction; how they were historically constituted; how their reproduction is ensured; how this structure of relations of production and productive forces itself structures the overall social formation—we find that, on all these points, Baran remains silent. Instead of attempts to analyse these problems, the concept of surplus appears simply as a convenient means for *describing* a number of the ways in which the surplus-product is used in Third World formations; it can only describe effects of the structure. As to the determinants of the structure itself, the concept is of little value, and we are forced back again to Baran's inadequate thesis of determinance by capitalist penetration, which blocked indigenous economic evolution. Thus, the concept of surplus may be said to indicate a number of uses of the surplus-product, but it does not provide a basis for explaining the existence of the surplus-product, nor why it is utilised in these particular ways in an underdeveloped social formation, nor what the effects of this utilisation are upon the overall social formation itself. As a result, there is no real basis here for a theory of underdevelopment.

Having assessed the effects of the basic concepts of Baran and Sweezy's theoretical framework on the analysis of underdevelopment, we can now state our conclusions, before going on to our examination of Frank's analysis.

BARAN'S THESES

All modes of production produce a surplus. We have indicated that the utilisation of this surplus-product is dependent upon the dominance of sets of relations of production, which differ from one mode of production to another. Thus, the particular forms of existence of Baran's surplus and its utilisation are themselves a function of the structure and reproduction of specific *modes of production*. Consequently, they are effects of the structure of the mode of production and thus differ from one mode to another. Therefore, to assert that an economic surplus is produced is inadequate, and to assert that its determinants (costs and output) remain constant is quite incorrect. Nevertheless, Baran and Sweezy take these assertions as an analytical basis for their theory of capitalist development and non-capitalist modes of production. We have tried to show how these assertions prevent them from analysing the structure of any particular mode of production, since they lead them to focus on effects rather than determinants. Herein lies their basic error.

If they had approached the structure, reproduction and development of the capitalist mode of production, they would have been able to situate consumption and the rhythms of consumption in relation to production and distribution; if they had done this, they would not have attributed all forms of capitalist penetration to one underconsumptionist cause, and they would have then been able to analytically demarcate the specificity of these stages, the fact that they result in different types of penetration of non-capitalist modes of production. They could then have begun to examine the effects of each of these stages for the development of capitalism in Third World economies.

Similarly, we have seen that the concept of economic surplus acts as a barrier to any analysis of the non-capitalist mode that co-existed with and pre-existed different forms of capitalist penetration. Due to the limitations of this concept, Baran is led to simply equate the transition to dominance by a capitalist mode of production with Marx's brief analysis of the development of capitalist production as it emerged in Western Europe, and more particularly in England, Marx's specific example of transition. There is no reason whatsoever to suppose that the particular mode of production which pre-existed capitalist development here also pre-existed capitalist penetration in all Third World societies. Rather, it seems that the elements of the capitalist structure can be constituted in many different ways, and within very different modes of production.

The absence of any real analysis of non-capitalist modes, combined with the absence of any analysis of different forms of capitalist penetration of these modes leads Baran to the inadequate conclusion that this penetration simply 'blocked' the evolution of capitalism. From this follows the corollary that economic development *per se* is predicated on a removal of this penetration. This argument then becomes the foundation for a simplistic opposition that reduces the theoretical analysis of an underdeveloped society to a description of how the surplus that is produced is utilised. Why and how this surplus is produced or utilised in these particular ways cannot be determined since the problem of the structure in which the surplus exists and is utilised is never posed.

FRANK: THE EXTENSION OF UNDERDEVELOPMENT

Turning now to Frank's texts,[17] we will try to show both how the particular limitations of his analysis are an effect of the concepts outlined above, and how, by extending the application of these concepts, he provides the foundation for a Sociology of Underdevelopment. As with the Sociology of Development, we will then examine the explanatory adequacy of this discourse in terms of its theoretical objective.

Frank's analysis begins by accepting Baran's formulation of the impossibility of capital accumulation within a closed capitalist system. From this, the conclusion then follows that the limitations of this 'closed' system establish a necessary structural foundation for all periods of capitalist penetration, which are characterised by the extraction of the underdeveloped countries' economic surplus by the metropolitan national social formation. Underdevelopment is thus analysed as being a historical product of the relations between the underdeveloped society (Frank terms it a 'satellite' of the 'metropolis') and the penetrative capitalist nation-state.

> Thus the metropolis expropriates economic surplus from its satellites and appropriates it for its own economic development. The satellites remain underdeveloped for lack of access to their own surplus . . .

and, again,

> One and the same historical process of the expansion and development of capitalism throughout the world has simultaneously generated—and continues to generate—both economic development and structural underdevelopment.[18]

According to Frank, contemporary underdeveloped formations were originally 'natural economies', but with capitalist penetration they rapidly became capitalist. Thus Frank argues that capitalism has been the dominant mode of production in Latin America since the sixteenth century. In doing this, he polemicises against all forms of Dualism, from Boeke to Lambert. An essential aspect of his argument here is that,

> . . . once a country or a people is converted into the satellite of an external capitalist metropolis, the exploitative metropolis-satellite structure quickly comes to organise and dominate the domestic economic, political and social life of that people. The contradictions of capitalism are recreated on the domestic level and come to generate tendencies toward development in the national metropolis and toward underdevelopment in its domestic satellites just as they do on the world level.[19]

Thus, the exploitative form of the relationship between metropolis and satellite previously outlined also emerges within the underdeveloped formation as a direct result of capitalist penetration.

Given these propositions, a number of corollaries or hypotheses can be logically deduced. Frank groups these under the heading of 'Continuity in Change'.[20] The development of the subordinate metropoles is limited by their satellite status, as is the development of the domestic satellites. These satellites experience their highest levels of economic development only when their relations with the industrial capitalist monopolies are severed—e.g. as a result of events such as the 1930s depression, the First World War, and so on. Finally, those areas which are contemporarily the most underdeveloped were those which were historically the most closely tied to the metropoles.

For the purpose of our critique, we can conveniently summarise Frank's thesis in the following way:

(i) Metropolitan capitalist social formations penetrate natural economies and appropriate their economic surplus.
(ii) This is the direct result of contradictions within the social formation. (These contradictions are unspecified, but the reference to them in Frank's texts indicates that they are problems generated by the familiar underconsumptionist tendency.)
(iii) This penetration almost immediately leads to the reproduction of capitalism as the dominant mode of production, through a destruction of the existing natural economy.
(iv) Capitalist penetration, through appropriating the indigenously produced surplus, promoted, and continues to promote a process of underdevelopment which is not only economic, but also political and social.[21]

THE LIMITATIONS OF FRANK'S BASIC CONCEPTS

Capitalism is clearly the most crucial concept of Frank's discourse; capitalism generates underdevelopment and, in so doing, creates capitalism. The immediate questions that arise are, therefore, what is this capitalism, why does it penetrates natural economies, and how does this process generate underdevelopment and further capitalism?

It seems to us that Frank avoids answering these questions that are crucial corner-stones of his analysis. This is clearly symptomatic, since—as we have argued—the discourse in which he is working prevents these answers being given. The concept of economic surplus, for example, precludes any rigorous analysis of the structure, reproduction and development of modes of production; hence it cannot provide an adequate basis for analysing either the development of capitalist penetration of non-capitalist modes, or the existence of different forms of this penetration. Frank's use of this concept thus results in his defining capitalism in the most general terms possible.[22] He speaks of: 'Capitalism's essential internal contradiction between the exploiting and exploited which appears within nations no less than between them.'[23] Elsewhere, he refers to 'the contradiction of monopolistic expropriation/appropriation of economic surplus in the capitalist system . . .'[24] Yet, despite the vagueness of this definition, it plays a crucial role in Frank's argument, since

> . . . [the] metropolis-periphery division rests on captalism's essential internal contradiction between exploiting and exploited and, in the present as in the past, simultaneously gives rise to economic development for some and underdevelopment for most.[25]

What can we conclude from these statements? It would seem that the auto-genesis of a presumed 'essential contradiction' is responsible for underdevelopment. But how? Apart from several allusions to Baran's underconsumptionist analysis and the situation of the necessity of capitalist penetration within this, there are no reasons actually given in Frank's texts for the very existence of capitalist penetration, let alone its different forms. Frank speaks of 'contradictions' that generate underdevelopment, of 'monopoly capitalist structure', of 'the historical process of the expansion and development of capitalism', all of which, however, remain largely undefined. Although the axiomatic problem for any theory of underdevelopment is, as Frank states, that of the 'rationale' for capitalist penetration, no rigorous answer is, in fact, given to this question in his work.

Similarly, the fact that different forms of capitalist penetration have very different penetrative effects remains unanalysed. This result is much more evident in Frank's work than in that of Baran. For Frank, all stages of capitalist development are conflated into a single process in which surplus is extracted and has to be absorbed, in which the different effects of different stages of penetration are either ignored or confused.

To exemplify this point, take Frank's analysis of 'The Development of Underdevelopment in Chile', where he argues that, from the sixteenth century, European capitalism began a process of external expansion which created capitalist modes of production by destroying 'natural economies', a process which has continued up till the present. This analysis contains a number of serious problems.

Firstly, how can we speak of European 'capitalism' when in the sixteenth century it is clear that the structural preconditions for capitalist production did not, as yet, exist? Its fundamental precondition—a co-existence of accumulated capital with workers selling their labour-power on the market—was clearly absent during this

period, which was characterised by the dominance of a feudal mode of production, whose reproduction requires a non-separation of the direct producer from his means of production.

Secondly, if capitalism was not the dominant mode of production during this period, how can it be argued that 'capitalist expansion' promoted 'capitalist underdevelopment'? During the sixteenth and seventeenth centuries in Western Europe, there did, of course, exist a commercial class which was able to accumulate capital through overseas trade, through the process of converting commodities into money by sale and of money into commodities by purchase, but this is in no way indicative of the existence of a dominant capitalist mode of production, since this form of commodity-exchange can (and did) exist prior to capitalist production itself. All the existence of a 'commercial class' requires is a process of commodity circulation—the limited development of a market manifested in the unification of direct producers through the medium of a merchant. The existence of merchants' capital in Europe did, of course, lead to a European entry into non-capitalist societies. As we shall see later, this largely took the form of an intervention in the production of use-values, to promote the production of exchange-values for the market. This resulted in the gradual domination of the process of production in non-capitalist modes by the requirements of merchants' capital. This domination generally either led to the introduction of a form of quasi-feudal land-ownership or to the reinforcement of pre-existing non-capitalist relations of production. But what has any of this to do with capitalism? The process of penetration does not depend upon capitalist production, nor does it in any way establish the preconditions for capitalist development, since this form of penetration—as we shall see later—rather than separating direct producers from their means of production, either introduces or reinforces a pre-existing non-separation.

Thirdly, what equivalence is there between his penetration under the dominance of merchants' capital, and, for example, penetration during the period of late nineteenth-century capitalism? As we shall see later in our analysis, in the latter case penetration resulted from the attainment of a specific stage in the development of capitalist production, which itself had effects on the non-capitalist mode that were particular to this stage. In the former case, capitalism obviously cannot produce analogous effects, since it is not even the dominant mode of production.

Citing these problems indicates how Frank's failure to define capitalism as a system of production whose reproduction and development require specific forms of capitalist penetration of non-capitalist societies, leads to major absences in his analysis. Apart from failing to provide an adequate theoretical basis for answering the question as to why the social formations of the Third World were historically subject to penetration by 'more advanced' modes of production and by capitalism in particular, Frank also fails to answer the all-important question for any theory of underdevelopment as to what were the different effects of different stages of capitalist penetration of non-capitalist modes. All he can reply is that they 'produced capitalism', which is not only inadequate, but quite incorrect, if by capitalism we mean a specific form of production.

These points can be further exemplified by examining in greater detail one of Frank's major contentions, namely that capitalism has been dominant in Latin America since the very beginning of European colonisation. This point is of considerable importance, since it establishes the basis for major conclusions on important Latin American formations. For example:

> The domestic, economic, political and social structure of Chile always was and still remains determined first and foremost by the fact and specific nature of its participation in the world capitalist system and by the influence of the latter in all aspects of Chilean life.[26]

What, then, is the basis for the conclusion that Latin America has been '. . . capitalist not only from its birth but from its conception'?[27]

In an article entitled 'Capitalist Latifundio Growth in Latin America',[28] Frank extends his definition of capitalism by stating that it also includes production for the market. Adding this to his previous formulations (above), we can conclude that by capitalism, Frank means a system in which the surplus product is appropriated by non-producers in a process of producing commodities for a market; and since these phenomena have been in existence in Latin America since the sixteenth century, then it follows that 'capitalism' has existed since that time, that there has constantly been and still is a dominant capitalist mode of production. This conclusion clearly constitutes a major error. Surplus labour is extracted and appropriated from direct producers in varying modes of production. Similarly, many different modes of production produce commodities for a market. Taking Frank's definition we would be forced to conclude, for example, that capitalism was dominant in Rome, Greece, the European 'middle ages', the slave plantations of America, etc., since a surplus was, of course, extracted from the direct producers, and commodities were produced for the market. Frank defines capitalism so loosely here that it can simply be equated with commodity-exchange and the circulation of commodities, since the concept of surplus that he operates with precludes any conception of capitalism as a mode of production. Furthermore, this error produces effects which seriously reduce the explanatory validity of Frank's underdevelopment theory.

In analysing any particular Third World society, we initially have to investigate the system of production present within it, prior to capitalist penetration. We then have to establish the effects that different forms of capitalist penetration of this mode had on its development, establishing to what extent these forms created the preconditions for the reproduction of capitalism as the dominant economic system. In all cases, we discover that, despite the increasing prevalence of capitalist production, elements of the pre-existing system continue to be reproduced. To take an example: in many contemporary Third World societies, the continuing non-separation of direct producers from their means of production within the agricultural sector acts as a barrier to the development of capitalist production. This restriction results from the continuing reproduction of the labour-process of a previously dominant system of production. For example, as a number of anthropologists have recently indicated in the case of societies in West-Central Africa,[29] the continuance of the labour-processes of the pre-existing system of production within a social formation in which capitalism is becoming increasingly dominant, results in a confining of potential capitalist labour-power within the tribal units of production. Again, for example, in Latin American societies, the landed property relations set up by Spanish colonialism require, for agricultural production, that direct producers retain possession of their means of production: these co-exist with capitalist relations of production which require a separation of direct producers from control and ownership of the means of production. Consequently, the landed property relations thus present real barriers to capitalist development. Here we have just one example of the effects of a combination of two systems of production. If we were to analyse other social phenomena—ideologies, the political structure, etc.—we would again find that they combine elements required for the reproduction of very different societal types. Posing in this extremely schematic form what, for us, must enter into the theory of the Third World formations, enables us to see more clearly the limitations of Frank's notion of capitalism.

His thesis that Latin America has been capitalist since the sixteenth century prevents us posing crucial questions in the analysis of Third

World formations, since it reduces the complexity of their structures to a simple determinacy by capitalist penetration (as Frank himself loosely defines it). His thesis that 'capitalism' simply creates 'capitalism' by destroying pre-existing 'natural economies', that this new capitalism is totally determined by the development of the old pre-existing 'capitalism' removes by fiat crucial problems in the analysis of Third World societies at the outset. Let us take one such problem: the continuing persistence in Latin American societies of economic and social phenomena produced by non-capitalist systems of production.

As we indicated above it seems that, in many Latin American countries, early capitalist penetration in the sixteenth century established landed-property relations which reinforced the already existing non-separation of producers from their means of production (this seems to me the best way, for instance, to characterise the Latin American hacienda). Furthermore, these relations (and the political, social and ideological formations that they give rise to)[30] continue to persist. Given this, if we then take Frank's assertion of a constant capitalist dominance, we would have no basis for analysing these semi-feudal relations of production, since the dominance of capitalism would deny their very existence. This, of course, is exactly what Frank does. In his essay, 'Capitalism and the Myth of Feudalism in Brazilian Agriculture' (which he posits 'may also find application elsewhere in Latin America and perhaps even in Asia and parts of Africa'[31]), he argues that,

> On the contrary, all of Brazil, however feudal-seeming its features, owes its formation and its present nature to the expansion and development of a single mercantalist-capitalist system embracing (the socialist countries today excepted) the world as a whole—including all of Brazil.[32]

His proof of this assertion is perhaps even more symptomatic, since the only way in which he can prove that semi-feudal relations of production never existed is to equate feudalism with non-market production and a closed non-monetary mode of production (as opposed to capitalism with its open, monetary, market-production) and to then state (as is, of course, 'correct') that capitalist penetration never produced this feudal reality. Such a conclusion can, of course, only exist as a function of his equating capitalism with commodity circulation, and the consequent impossibility of establishing feudalism as a 'closed' system of production. This, however, provides no proof that capitalist penetration did not establish landed-property relations with effects similar to those of feudal relations of production, since these very property relations can (and, in fact, in most feudal modes of production did) exist within market production, monetary circulation, and the somewhat less serious indicators of capitalism that Frank gives, namely, 'a certain liberty', 'specialisation', and 'land conservation and maintenance'.[33]

Consequently, Frank's thesis that capitalist penetration has produced capitalism sets up—as can be seen from this example—a real barrier to analysing an important result of early capitalist economic penetration of contemporary Third World societies (and most notably Latin America). Not only does his analysis conceal this result, but the concepts with which he operates lead him to deny its very existence.

From this brief examination of the way in which the basic notions of underdevelopment operate in Frank's texts, we can conclude that they appear to be inadequate for analysing their object, and that they can conceal fundamental problems in the analysis of Third World formations. Furthermore, they also tend to reduce explanation of all aspects of the social structure to a form of economic determinism. Since, in our view, it is this reduction which founds the possibility for the Sociology of Underdevelopment, it is essential that we now look further into this determinism and its effects on the discourse of Underdevelopment.

THE SOCIOLOGY OF UNDERDEVELOPMENT

In our analysis of the theory of the Sociology of Development, we concluded that, by abstracting concepts from existing discourses, it had been able to construct a system in which Third World societies were analysed as traditional forms between two hypothetical states of equilibrium. A particular social phenomenon is explained in terms of the extent to which the subject operating within the theory perceives it as contributing to the future attainment of the desired state. The basis for the explanation of a social phenomenon within this schema is, as we argued, inherently teleological. Given this, as we attempted to show, the discourse is incapable of explaining future possible directions of change, or of analysing Third World societies, except within limited teleological terms. The substitution of a particular model of an advanced capitalist social formation for the equilibrium state, and the substitution of the subject's opinion as to the degree to which an existing phenomenon can contribute to the attainment of this state for the real concept of function within the discourse produces a reductionism, in which attempts to rigorously explain the structure and reproduction of the Third World formation are seriously restricted.

Given the limitations of this theoretical framework, the work of Baran, Sweezy and Frank has indicated a way forward. The great value of their work is that, by operating within their discourse, the limitations of the Sociology of Development and its value-laden basis are clarified, and the *bête noire* of the subject—the fact that capitalist penetration of non-capitalist modes has placed and continues to place major restrictions on the latter's development (together with the social and political development of the societies in which these modes exist)—is made absolutely clear. However, clarity in direction, indicating the site for theoretical work, is merely an initial step, and it now seems to me that in the present situation the work of Baran, Sweezy and Frank has come to be a barrier to further progress in the analysis of Third World societies. Furthermore, although their conclusions induce an awareness of the limitations of the Sociology of Development, they nevertheless operate in a discourse that is in many ways similar to the one they are criticising.

In their analyses of underdevelopment, Baran and Frank establish two states of existence of the surplus and its mode of utilisation. One is characteristic of stages prior to and during capitalist penetration, when the surplus is utilised through appropriation by foreign interests, military establishments, excess consumption, etc.[34] Another is a potential state, where this appropriated surplus could potentially be utilised (in what would be 'a more rationally ordered society'[35]) to meet the present and future internal needs of the indigenous population. It is this second state which provides the basis for the concept of *under*development in the BSF discourse. Thus Baran states: ' . . . in the underdeveloped countries, the gap between the actual and the possible is glaring, and its implications are catastrophic.'[36]

Baran is much more specific than Frank as to exactly what he means by this rationality ordered society (if only at the economic level). Its basis is the 'mobilisation' of the potential surplus (viz. 'the expropriation of foreign and domestic capitalists and landowners, and the consequent elimination of the drain on current income resulting from excess consumption, capital removals abroad and the like', producing 'an instantaneous increase of the actual economic surplus'[37]), the reallocation of unproductive labour, the planned development of agriculture related to industry on the basis of the new mobilisation of the surplus (a process which is determined by the present and future needs of the indigenous population themselves), and so on. All these problems are analysed by Baran in the concluding chapter of *The Political Economy of Growth* where he shows how the utilisation of the potential surplus could result in a more rational organisation of a Third World society's surplus-product on the basis of a planned economic surplus.

The major point to be directed at this analysis is that—despite the total divergence of its conclusions from those of the Sociology of Development—it still constitutes a form of explanation in which the contemporary phenomena of an underdeveloped society are defined by being juxtaposed against a potential state, the achievement of which they do, or do not contribute to. In itself, it cannot provide a basis for explaining the existence of those phenomena that do not contribute to the attainment of this state. Consequently, the contemporarily given cannot be fully explained. All that can be done is that *the limitations of this given in relation to the potential can be pointed out*. When it goes beyond this level and tries to explain the 'non-contributory' phenomena—i.e. to explain the present situation—the BSF discourse resorts to a reductionism in which all those phenomena that contribute to the preservation of the present state are analysed as being the effects of a single cause, namely 'capitalist penetration', which itself arises from the impossibility of surplus absorption. Thus, we initially have phenomena explained teleologically, in terms of the extent to which they do, or do not, tend towards the future achievement of a potential state. Those phenomena which, on the basis of the above, are defined residually as tending to preserve the existing state, are then simply reduced to the effect of one causal factor, to being an expression of the latter. Consequently, the basis for a definition of the present appears to be a hypothetical potentiality. Yet, most phenomena do not contribute to the achievement of this potentiality, and these are then explained as being the effects of a single cause. The analysis of underdevelopment is only able to operate given these teleological and determinist axioms.

The teleological axioms seem to me to be subject to the same criticisms made with regard to the Sociology of Development, the crucial difference being that the state postulated by Baran and Frank can indicate the existence of capitalist penetration and describe its operations. Nevertheless, the present is still defined in terms of the potentiality of a future state, and the extent to which elements of the present could function to bring this about. Even though this potential state is, to a certain extent, made more concrete in Baran's work by a brief analysis of economic developments in one society—the Soviet Union prior to the fifties—which, by appropriation and internal investment of its surplus, was able to break out of the underdeveloped state, this does not provide an adequate basis for analysing contemporary underdeveloped societies. To juxtapose an existing against a potential state does not constitute an explanation of the former's existence, but merely indicates its limitations. So, in order to explain the contemporarily existent more adequately, the BSF discourse has to have recourse to its *reductionist elements*.

Here all phenomena that cannot contribute to achieving the potential state are explained by analysing them as a simple effect of the results of capitalist penetration itself. Frank and Baran largely confine their analysis of this effect to the economic level, but there clearly is a basis for extending this to other phenomena. Indeed, Frank himself makes the following statement in his analysis of Chile:

> The domestic *economic, political and social* structure of Chile *always* was and still remains *determined first and foremost* by the fact and specific nature of its participation in the world capitalist system, and by the influence of the latter on all aspects of Chilean life.[38] [our emphasis]

or, again,

> Underdevelopment in Brazil, as elsewhere, is the result of capitalist development.[39]

Thus, other than economic phenomena can similarly be explained (in a relatively more or less simple or complex manner) in such reductionist terms. It is this which lays the foundation for a Sociology of Underdevelopment.[40] This approach has serious limitations for the analysis of Third World formations. For the purpose of our critique, these can be sub-divided into three aspects.

Firstly, there are crucial phenomena which must remain unexplained, or, at best, inadequately explained if we remain within the framework of the Sociology of Underdevelopment. We gave the example above of the importance of analysing the economic effects of semi-feudal relations of production which continue to be reproduced alongside capitalist relations in some Third World formations. We could equally well have cited many others which are the direct result of the survival of non-capitalist forms within Third World societies. For example: the contradictory political alliance between the sometimes mutual, sometimes opposed interests of capital and landed property in the transition to dominance by a capitalist mode of production; the articulation of capitalist ideologies in ideological forms appropriate to non-capitalist societies; the survival of kinship structures concomitant with minor changes in their mode of functioning which, resulting from capitalist penetration, ensures the provision of a temporary labour-force which is so vital in the early stages of penetration. Because there is no basis within the underdevelopment discourse for analysing the origins or development of such phenomena, which result from a continuing combination of capitalist and non-capitalist forms, they can only remain either symptomatically absent, or inadequately explained as being simply the results of capitalist penetration.

Secondly, it is argued that any process of internal surplus utilisation in production for the domestic market has *one major determinant*, namely a decline in the intensity of capitalist penetration. Conversely, as long as this is pervasive, then such a process of internal development must remain absent, and the social formation will be underdeveloped. This argument is basic to Frank's thesis. Referring to Brazil, for example, he concludes:

> What happened in the colony was *determined* by its ties with the metropolis and by the intrinsic nature of the capitalist system. It was not isolation but integration which created the reality of Brazilian underdevelopment . . . What economic development did take place occurred in places and times in which the ties with the metropolis were less close.[41]

At the economic level, this argument is unsatisfactory, since it simply reduces the process of internal surplus-utilisation for the domestic market to non-satellite status. Its validity can be exemplified by briefly looking at one case Frank chooses—that of Japanese industrialisation.[42] Here he argues that the limited development of capitalist penetration of Japan in the nineteenth century enabled it to avoid the metropolitan-satellite relationship, and to invest its surplus internally, which for Frank seems to be equated with industrialisation. This conclusion is highly questionable. In the early years of Japan opening up its trade, it was able to establish a monopoly over the world silk trade, due to crises in the previously dominant European sericulture. However, by the late sixties, trade deficits began to develop, and the high rates of capital accumulation previously attained began to fall. Now, if Frank's argument is correct, then Japanese capitalism should obviously have developed by this stage, since no capitalist mode of production had penetrated, and a large surplus had been accumulated which could have been invested internally largely through importing available industrial goods. However, we know very well that Japanese capitalism failed to develop until after 1868, following the Meiji restoration. Frank's schema is limited here in explaining the specific determinants of capitalist industrialisation, primarily, of course, because (as we have seen) he cannot *establish the structure within which the surplus was extracted*, and *the dominant relations of production within this structure*. When we know this, we can see that the Meiji restoration of 1868 produced profound transformations in the relations of production which then established a basis for capitalist development—that prior to the Meiji restoration the surplus took two forms—money capital in the merchants' capital sector and ground rent in the agricultural sector. These forms were then transformed by the restoration, the appropriation of ground-rent by the government enabling the latter to be transformed into industrial capital.[43] Given Frank's schema, none of this could (or would have to) be known, since the possibility for industrialisation/internal investment of the surplus is simply (and incorrectly) reduced to an absence of capitalist penetration, to a single causal factor.

We can reach similar conclusions with regard to Frank's thesis that the industrial development of the satellite countries in the twentieth century is inversely related to their contacts with the metropolitan countries. One of the most notable cases cited by Frank of indigenous industrialisation during this period in Latin America occurred in Peronist Argentina largely *after* the re-establishment of ties with the metropolitan countries. Furthermore, this strategy failed not simply because of 'increasing economic contact' with 'the metropolis', but primarily due to the contradictory political alliances on which the industrialising strategy was based, and the constraint's which the latter imposed on both the landed oligarchy and the working-class. Here we see again (and other examples could have been chosen[44]) that the incidence of indigenous industrial development directed towards the domestic market is in no way simply primarily reducible to an 'absence of capitalist or imperialist penetration'.

Despite these limitations, however, it is clear that, in the same way that Frank carries out such an economic reductionism, the basis is also laid here for a concomitant social reductionism, in which the 'lack' of 'social development' is simply explained by it being reduced to an effect of a conflated capitalist penetration. Indeed, Frank himself states,

> . . . once a country or a people is converted into the satellite of an

> external capitalist metropolis, the exploitative metropolis-satellite structure quickly comes to organise and *dominate the domestic economic, political and social life of the people*. The contradictions of capitalism are recreated on the domestic level, and come to generate tendencies toward development in the national metropolis and toward underdevelopment in its domestic satellites.[45]

or again,

> Therefore, the *economic, political, social and cultural institutions and relations* we now observe there (in the 'underdeveloped world') *are the products of the historical development of the capitalist system* no less than are the seemingly more modern or capitalist features of the national metropolis of these underdeveloped countries.[46]

I do not really want to develop this point any further, since its implications are clear. In analysing Third World societies, the Sociology of Underdevelopment uses a unilinear social determinism, in which all non-economic elements are analysed simply as effects of a capitalist penetration which universally blocks the development of the pre-existing society. This inversion of the object of the discourse of the Sociology of Development places severe restrictions on the analysis of any phenomena within the reality that it purports to explain.

The third and final corollary of the above tenets is, of course, that the sole precondition for indigenous economic and social development in any Third World formation is a removal of the barriers created by capitalist penetration.

> Therefore, short of liberation from this capitalist structure or the dissolution of the world capitalist system as a whole, the capitalist satellite countries, regions, localities and sectors are condemned to underdevelopment.[47]

This, it seems to me, is where Frank's thesis—and the Sociology of Underdevelopment in general—is at its weakest.

We analysed earlier how the Sociology of Development itself explained phenomena teleologically as indices which could function for the achievement of a particular equilibrium state. The Sociology of Underdevelopment can now clearly be seen as an inversion of this discourse. Its conclusion that indigenous development is attendant upon the removal of capitalist penetration can only be premised upon the claim that the latter does not bring about the potential state, and that it fails to do this because it does not promote what (from within the potential state) can be defined as 'indicators' contributing to this potentiality. On this basis, it can then be argued that all that has to be done is to remove this penetration. Such a conclusion is ultimately premised on a comparison of the given with the potential, and the further assertion that capitalism alone retains the given. As we have seen, the first premise is no more than a useful pointer to the presence and effects of a conflated capitalist penetration, and the second is grossly inadequate. Therefore, the conclusion that underdevelopment will simply be alleviated by the removal of capitalist penetration cannot be established on the basis of these premises. But, more than this, we must conclude that the assertion that the simple 'removal' of penetration could, in some unspecified way, 'create' development (defined in terms of a given potentiality) is ultimately as inadequately established as the assertion that implanting further elements of this penetration will create a basis for indigenous development (defined in terms of attaining a given state of equilibrium). Neither can in any way act as a substitute for a rigorous analysis of the structure of Third World formations, or the future directions of change possible from within the tendential development of their structure.

Part II: Towards a Historical Materialist Theory of Third World Formations

4 Theoretical Prerequisites for an Analysis of Third World Formations: Theses

Having set out a critique of the two discourses that claim to provide an explanatory framework for analysing Third World societies, showing how they provide inadequate bases for explaining their object of investigation, we can now move on to the main object of our text.

As an introduction to this—even at the risk of introducing concepts with which the reader is not yet familiar—we will briefly outline here a set of theses, which will act as guidelines for our analysis of the prerequisites for theoretical work on what have formerly been termed 'underdeveloped' or 'modernising' societies.

1. At the outset, it is necessary to reject the notion of underdevelopment. As we argued in Chapter 3, this notion cannot provide us with any rigorous basis for analysing the existence, forms, or effects of the various types of capitalist penetration within societies dominated by non-capitalist modes of production. As we have seen, its teleological and economistic basis largely precludes it from approaching fundamental problems in the analysis of the structure and historical development of these societies.

It seems to us that analysis of those societies that have previously been designated by these notions can only be rigorously approached theoretically from within the discourse of historical materialism. This, of course, is the site of a real problem, since very little work has actually been done in this area, and we will have to undertake a detailed reading of particular aspects of the discourse of historical materialism in order to achieve our object.

2. As opposed to the notion of underdevelopment, we will argue that the contemporary reality it alludes to must be analysed from within historical materialism as *a social formation which is dominated by an articulation of (at least) two modes of production*—a capitalist and a non-capitalist mode—in which the former is, or is becoming, increas-

ingly dominant over the other. It is this articulation which, as we shall see, contemporarily structures the social formations of the Third World. Consequently, it is of fundamental importance to establish exactly what this articulation is and how it is reproduced. But, beyond this, in order to examine the history of these social formations we must also begin to analyse the *preconditions for the existence of this articulation of modes of production.*

3. In order to analyse the preconditions for the emergence of a capitalist mode of production, of how they come to be articulated within a social formation dominated by a non-capitalist mode, and how the latter is increasingly subordinated to the former, we must initially approach the problem of what we can term the genealogy of the capitalist mode, of how its elements were historically constituted. Within the Third World formation, the historical field of this genealogy is that of the development of a social formation in which a non-capitalist mode is dominant, and within which various forms of capitalist penetration intervene. So, in order to begin to analyse this historical field, we must initially establish:

(i) *The structure of the mode of production that was dominant within this social formation*, together with the *reproduction* of this mode and the *form of development specific to it.*

(ii) The possibility for different forms of capitalist penetration of non-capitalist modes from within the development of the capitalist mode of production, analysing how these different forms were articulated in the development of particular non-capitalist modes of production, and what were the specific effects of these different forms, both upon the dominant mode of production and within the development of the overall social formation. Finally, we must analyse the extent to which these different forms establish the pre-conditions for the existence of the particular combination of elements that constitutes the capitalist mode of production.

It is only by beginning to answer these two problems that we can establish *how a capitalist mode of production comes to exist and to subordinate the previously dominant non-capitalist mode to its own increasing dominance*, i.e. that we can establish the historical preconditions for the articulation of modes of production that structures the contemporary social formations of the Third World.

4. In general terms, then, the reality that has previously been approached via the notions of 'underdevelopment' or 'modernising society' will be analysed here from within historical materialism as a *social formation that is transitional*. The specificity of this transition lies

in it being *brought about largely by capitalist penetration, and more particularly by one of its forms, imperialist penetration*, which, as we shall see, has as its *specific economic effect a separation of direct producers from their means of production.* The Third World formation can, therefore, be analysed as being structured (or determined in the last instance) by an articulation which is produced largely as an effect of imperialist penetration.

Once we have analysed this, we then have a rigorous basis for examining the complex historical development of the different levels of the social formation during this period, and, consequently, we have an initial basis for analysing particular contemporary Third World formations, and their future possible directions of change.

5. As a result of its being determined in the last instance by an articulation of modes of production, the Third World formation is characterised by a whole series of *dislocations*[1] between the various levels of the social formation. As opposed to the previous period of determinancy in the last instance by a particular non-capitalist mode, in which the different levels were *adapted to one another*, the latter are now *dislocated with respect to each other, and with respect to the existing economic structure itself.* Imperialist penetration intervenes economically, politically and ideologically within these dislocated levels in order to ensure the increasing dominance of the capitalist mode of production, and to create that restricted and uneven form of economic development (together with its political and ideological guarantees) we described in Chapter 2. Yet, it is also the case that the existence of these dislocations, and the effects that imperialist penetration has upon them in trying, as it were, to adapt them to the political and ideological reproductive requirements of a capitalist mode of production, can produce—in specific conjunctures[2] in the transition—the possibility for the emergence of the preconditions for the constitution of a different mode—a socialist mode—of production. Thus there can be no such thing as a 'linear succession' from dominance by a non-capitalist to dominance by a capitalist mode of production. Imperialist penetration, having as its object to create the preconditions for a transition to a specific form of capitalist production can produce—as we shall see later—the preconditions for the possibility of a socialist mode of production.

These, then, are the major theses which will govern our theoretical analyses in the remaining chapters. The raw material on which we will work is formed largely from particular texts written from within the discourse of historical materialism that approach the problems we are posing.

Having elaborated very schematically the theses that will guide our analysis, we must now turn to the first step in this, by specifying as succinctly as possible the theoretical object of the discourse of historical materialism—the social formation, its structure, reproduction and dynamics.

5 Social Formation and Mode of Production

SECTION 1: SOCIAL FORMATION

Moving on to our main objective, we can now pose the question of the theoretical prerequisites for the analysis of Third World formations. This requires a preliminary note.

THE CONCEPT OF SOCIETY

Whether Third World societies are analysed from within a framework based on modernisation or underdevelopment theory, they are always referred to as 'societies'. The fact that the same term is used conceals a crucial theoretical point which is often neglected. For example, the Parsonian discourse has as its object 'society', but this can only be conceived within the discourse as a structure whose elements are unified as realisations of the functional essence of the structure, realisations which can only operate given the existence within the discourse of other notions—equilibrium, integration, etc.,[1] notions which preclude any explanation of change beyond *ex-post-facto* generalisation. Similarly, the Weberian totality has a specific conception of 'society', in which the structural elements are unified by and ultimately reduced to being realisations of the historical emergence and development of the phenomenon of rationalisation.[2] Again, the notion of society in the texts of Baran, Sweezy and Frank is necessarily accompanied by the teleological assumptions inherent to their discourse through their use of the concept 'economic surplus and its utilisation'. 'Society' as it is used in all these analyses is never a notion that is given, but is always inscribed in a discourse whose theoretical object, and concepts corresponding to this object, determine and limit its explanatory adequacy. This discourse—which has its own conditions of existence[3]—structures the questions that will be posed within it to any aspect of reality that it is attempting to explain. The discourse defines what is essential and non-essential in the analysis of any phenomenon, and this is revealed by the problems that are absent and present within it. When any sociological analysis utilises the notion 'society', the discourse from which it is extracted necessarily embodies its presuppositions (even if these exist as an eclectic combination of concepts extracted from the texts of major theorists); these then determine and limit the problems posed by the phenomenon under examination.

For our analysis, this basic point cannot be overstressed, since, as we have seen, texts written in the Sociology of Development or Underdevelopment are firmly situated either within a Parsonian structural-functionalist discourse (adding elements extracted from other discourses) or are inscribed in a discourse centred on the notion of economic surplus. Hence, any use of the concepts 'modernising' or 'underdeveloped' society must remain restricted by the inherent limitations of these discourses. For this reason, the general notion of society cannot be accepted as given for our analysis; rather, it must be constructed from within the particular theoretical discourse used. Since our analysis will be carried out from within historical materialism, we need to clearly distinguish its theoretical object from those already outlined.

THE OBJECT OF HISTORICAL MATERIALISM: THE SOCIAL FORMATION

The concept of social formation is defined only in very general terms by Marx as comprising an economic structure (the mode of production), which he claims ultimately determines two 'superstructures'—the state and law, and ideology.

Recently—as is well known[4]—Althusser has developed this notion, arguing that the social formation exists as a complex totality containing a number of different practices—economic, political, ideological and theoretical—whose unity constitutes what he terms 'social practice'. Despite their particularity, each of these practices has a common invariant structure, in which a determinate raw material is transformed into a specific product; this transformation being effected by a labour which utilises particular means of production. Thus, economic practice requires a transformation of nature into social products through the methodologically organised use of means of human labour (means of production); ideological practice exists as a transformation of subjects' lived relations to the world through means of ideological struggle;

political practice as a transformation of existing social relations into new social relations; and theoretical practice as the transformation of concepts existing within ideological discourses into a specific product, scientific knowledge, by means of an abstract labour which works on concepts that ideologically define the field in which the problems of science are initially posed. Within the unity of practices in a social formation, economic practice is ultimately determinant, and structures the inter-relation of the various practices. Developing Althusser's reading of Marx on this point, it is essential that we now examine the component practices of the social formation in greater detail.

THE STRUCTURE OF THE SOCIAL FORMATION

Economic practice

In the *Contribution to a Critique of Political Economy*, Marx argues that:

> All periods of production, however, have certain features in common; they have certain common categories. *Production in general* is an abstraction but a sensible abstraction in so far as it actually emphasises and defines the common aspects and thus avoids repetition.[5]

These 'common categories' are specified by Marx in Volume 2 of *Capital*:

> Whatever the social form of production, labourers and means of production always remain factors of it. But in a state of separation from each other either of these factors can be such only potentially. For production to go on at all they must unite. The specific manner in which this union is accomplished distinguishes the different economic epochs of the structure of society from one another. In the present case (capitalist production), the separation of the free worker from his means of production is the starting-point given, and we have seen how and under what conditions these two elements are united in the hands of the capitalist, namely as the productive mode of existence of his capital.[6]

It seems, therefore, that—at its most basic level—economic practice, 'production in general', contains certain elements—the worker, his means of production, and the object on which he works; there also exists a non-productive worker who appropriates the worker's surplus labour. For production to take place, these elements must be combined.

Marx argues that they are combined through two different types of relations or connections, and that the manner in which they combine distinguishes different periods or 'modes of production' from each other. What then are these two connections?

First, in any one form of production, these elements combine together in definite *labour-processes*[7] by means of which the 'appropriation of nature', as Marx terms it, is organised; here raw materials—or other inputs—are transformed into products that have a use-value. This transformation requires that production be divided into a series of definite functions, with specific forms of co-operation between them, in order that the object (use-value) can be produced.

The elements contained within the labour process can, of course, combine in many different ways. Within capitalist production, for example, there can exist very different processes. There can be a simple co-ordination in one unit of tasks which were previously separated, the co-ordination being organised by an owner of capital who has the power to select a particular use-value for production. Alternatively a more complex co-ordination can develop in which the tasks set up in simple co-ordination become centralised, and then decomposed and re-organised to form a more complex division of labour in which productivity can be increased through a constant repetition of basic tasks. This reorganisation can then permit the introduction of machinery to replace the tools used by human labour, with a further fragmentation and hierarchical co-ordination being accompanied here by an increasing loss of control by the worker over the means of production he operates. Again, manual involvement by the worker in the labour-process can be eliminated through the use of electronic techniques, reducing the worker's role to one of control. Here the separation of the worker from both the means and object of production is total, as compared, for example, with simple co-ordination, where there is still a direct relation between the worker and his means of production, separated from his object of labour.

Whilst several labour processes, such as these, can co-exist in a particular form of production, they are, however, always structured by a set of relations of production. This brings us on to our second connection between the basic elements above.

The surplus labour existing in the labour processes as a result of production going beyond the simple reproductive requirements of the labour utilised in them produces a surplus-product which is subject to different forms of extraction in different periods. These forms are the result of definite *relations of production* which, once established, set up a system of differential access to ownership and control of the means of production for the worker and non-productive worker. The relations of production establish *one dominant form of extraction of surplus labour* for a specific mode of production. This extraction of surplus labour has primacy over and structures the labour processes.

We can see, therefore, that a particular mode of production exists as a specific double combination of different elements, forming a particular combination of relations of production and labour processes, structured by the dominance of the relations of production.

Thus, from 'production in general' with its 'common features', we can begin to move on to consider production in particular, to an analysis of a definite mode of production, which Balibar has aptly termed

> a system of forms which represents *one state of the variation* of the set of elements which necessarily enter into the process considered.[8]

Capitalist production, for example, can now be defined succinctly—at its most basic level—as a form in which the relations of extraction of surplus labour are such that the non-productive worker—the capitalist who effectively owns[9] and controls the use of the means of production—is able to set the worker to work on these means of production, as a result of the separation of the former from both ownership and control over the latter. The capitalist (the non-productive worker) purchases the worker's labour-power in exchange for a monetary wage, with which the workers buys commodities for personal consumption, only to have to re-sell his labour-power to ensure a further supply of the commodities essential for the reproduction of his labour-power. The capitalist on the other hand purchases means of production and commodities for personal consumption from other capitalist units of production. The surplus labour that provides the basis for these purchases and for capital accumulation can be appropriated by the capitalist owner of the means of production, since within the labour process there exists a difference between the value of labour-power and the new value that can be created by the application of this labour power to means of production. The surplus labour time not devoted to the reproduction of the value of the means of production takes the form of surplus labour, which the capitalist is able to appropriate through his ownership and control of the means of production. Thus, the specifically capitalist relations of production between worker, object, means and non-worker define a distribution of the means of production in which the latter are effectively owned and controlled by non-productive workers, with the worker having to sell his labour-power as a commodity to the capitalist. These relations of production require labour-processes in which the amount of surplus labour (in the form of surplus value) can be constantly increased (due to the competition between units of production) to meet the needs of enlarged reproduction, either through such means as extending the duration of working time and the intensity of work or by increasing the productivity of labour.

This combination, which specifies the capitalist mode of production, presents a different state of variation of elements from that found, for example, within the feudal mode of production. Here, whilst the relations of extraction of surplus labour (through a levying of ground-rent) involve the existence of a separation of producers from non-producers, and the right of the latter to portions of the surplus product, the workers who produce the product remain in possession of their means of production. Whilst in the capitalist process of production there was a coincidence between the various labour processes (the process of 'real appropriation') and the relations of extraction of surplus labour, this is clearly not the case in the feudal mode of production. To be more explicit, in terms of the relations of production in the capitalist process, the capitalist controls the means of production and the product produced, and the worker is separated from ownership and control. Similarly, in the labour process, he actually possesses the means of production and product, and the worker possesses only his own labour-power; relations of extraction and real appropriation, therefore, coincide. By contrast, in the feudal process, whilst the relations of extraction permit the non-productive worker to extract surplus labour from the direct producers, things are very different in the labour-process, where the producer remains in possession of his means of production and he himself (or in co-operation with other producers) produces the product. Thus, the non-productive worker's right to extract surplus labour is, as it were, 'contradicted' by what occurs in the labour process, since relations of extraction and possession do not coincide under feudal production.

This coincidence of the connections in capitalist production and their non-coincidence in other forms of production, such as feudalism, is of crucial importance since it *determines which of the practices in the social formation occupies the determinant place* and, therefore, which of these practices *structures the social formation within which each mode of production exists.* In those modes of production in which there is a disjunction or non-correspondence between the two connections, surplus labour can only be extorted by other than economic means, and, consequently, in social formations in which these modes are dominant, the latter determine an articulation of the practices in the structure in which *practices other than the economic itself are determinant.* Conversely, in those modes of production in which there is a coincidence between the connections, surplus labour can be extracted by economic means alone, and here the mode of production determines an articulation in which *economic practice is itself determinant.*

In the feudal mode of production, for example, the dislocation or non-coincidence between the labour processes and relations of surplus-appropriation means that the surplus labour cannot be appropriated simply by mechanisms existing within the mode of production itself. Marx indicates this when writing of the feudal mode:

> Under such conditions, the surplus-labour for the nominal owner of the land can only be extorted from them (the direct producers) by other than economic pressure.[10]

Consequently, in this mode of production, the non-coincidence of the two processes has as its immediate effect that the economic cannot be determinant on its own, since the extraction of surplus labour (which takes the form of levying of ground-rent on the direct producers) requires the intervention of a practice other than the economic within the process of production. As we know, in the feudal mode, this practice is the political, which, intervening within economic practice, determines for the land-owner a legally-defined right to a portion of the surplus product. Hence, the structure and reproduction of the feudal mode, in its generality (since we are not discussing here the variant forms of extraction of ground rent that can exist within it), requires a political intervention within economic practice itself. In other words, the structure of the feudal mode, existing as a definite combination of invariant elements, determines a particular articulation in which the political instance[11] occupies the determinant place within the structure of the social formation. As a result, it is the political instance that structures the inter-relation of the various practices within a social formation in which the feudal mode is dominant over other modes of production.

As a further example, take the lineage mode of production, whose presence a number of anthropological theorists have analysed in the pre-colonial social formations of West-Central Africa.[12] In this mode, the conditions of real appropriation are formed by family units which all belong to a specific lineage. The surplus labour produced in these units is however, extracted by the elders[13] of the different lineage groups in the form of élite goods (non-perishable goods, such as ironware, gold and leather products). In exchange of these élite goods, the elders supply women from units of the lineage other than those providing the goods; they also supply slaves from other lineages. Through these exchanges, the elders control the process of reproduction of the units by ensuring that the supply of women and slaves is such that a demographic balance is maintained between the provision of labour-power, non-productive labour, and consumption requirements. The maintenance of this balance is an essential condition for the reproduction of the conditions for production in social formations dominated by the lineage mode, and is ensured through an exchange of élite goods (produced in surplus labour-time) with the lineage elders. Here, therefore, as in the feudal mode, there is a dislocation between the relations of real appropriation on the one hand, and the relations of production under which surplus labour is extracted and which ensure the reproduction of the lineage mode on the other. This dislocation has as its immediate effect that the reproduction of the lineage mode cannot be ensured simply through economic practice, since the extraction of surplus labour requires an intervention within the economic which is both political and ideological, expressed in the determinance of production by kinship relations. Thus, in the lineage mode, the economic instance determines an articulation of the structure in which the determinant place is occupied by an inter-relation of the political and ideological practices in the social formation.

As opposed to the examples that we have given of the feudal and lineage modes, in which there exists a dislocation between the relations of real appropriation and the relations of extraction of the surplus product, within the capitalist mode of production there is a direct correspondence between these two connections. Within the labour process, the direct producer is separated from the means of production that he operates. Similarly, the relations of extraction of the surplus product (the relations of production) establish the ownership and control of the means of production by the non-productive worker (the capitalist), and the non-ownership and control of the worker. The economic instance thus produces here a determinancy in which it itself is determinant in the social formation.

From these examples, we can see that economic practice is determinant in the last instance, in that it ultimately determines which of the practices (or combination of practices) occupies the determinant place within a social formation dominated by a particular mode of production; it thus determines which practice structures the other practices in the reproduction of the social formation.

Having analysed the structure of production as a combination of labour processes and relations of extraction of surplus labour (relations of production), and indicated how the reproduction of the structure requires a determinacy by practices which varies from one mode to another, we can now extend our brief account of the mode of production by indicating how a particular process of production determines the emergence of specific forms of distribution, consumption and exchange.

In any mode of production, the relations of production govern both the distribution of income and the distribution of means of production and consumption. As a brief example of this, take the forms of distribution specific to a capitalist mode of production.

Here, as we have noted, the direct producer is separated from his means of production, the latter being effectively owned and controlled by the non-productive worker (the capitalist or agents of capitalist ownership and control) who is able to extract surplus labour under the form of surplus value, reproduce the value of the constant capital used up in the production process, and pay the wages of labour-power expended in production. The new value created in this process of production is distributed amongst the various owners of the factors of production. Thus, the labourer receives the value of his labour-power in a monetary form, and the surplus value is distributed to the owner(s) of the means of production and to those who have a legal claim to a portion of this surplus value (in the form of interest, rent, taxation by the state, etc.).[14] The relations of distribution of the value newly produced are, therefore, a function of the reproduction of the relations of production. Yet, distribution within the capitalist mode of production is not simply restricted to the distribution of the value newly produced. It is also concerned with the distribution of the products (or use-values) produced by the production process, whether these are use-values produced as means of production or consumption. The use-values produced for personal consumption are exchanged for the newly produced value distributed to the owners of the factor of production. This exchange is, therefore, dependent upon the distribution described above. Secondly, the use-values produced for productive consumption as means of production are exchanged amongst the owners of the means of production which is, again, a function of the process of production itself. We can conclude, therefore, that both types of distribution – of income and of the use-values produced for either productive or personal consumption – exist as functions of the process of production and the relations of production.

(Obviously, the distribution of the newly produced value in the capitalist mode of production is much more complex than this simplified example I have constructed for explanatory purposes. Surplus value is also distributed amongst those agents who ensure the circulation of commodities through their distribution, marketing and sale; the banking system has a claim on the surplus value through its function of ensuring the circulation of monetary capital; the state intervenes to abstract surplus value for its own use, etc. These points—and others—are ignored in our 'simplified' example, since we merely wanted to indicate here how distribution is—at the most basic level of reproduction—fundamentally a function of production relations. Thus, we are concerned with how the distribution of the value produced solely amongst the various owners of the factor of production is necessarily dependent on the relations of production.)

The structure of production similarly governs *consumption*, both productive and individual. Taking the capitalist mode of production as our example again, individual consumption, or the economic needs of individuals that are satisfied by consuming particular use-values, are dependent both upon the individual's disposable income and upon the nature of the products available. As we have seen, income is a function of distribution, itself dependent upon capitalist relations of production. Similarly, the availability of products is dependent on the level of development of the labour-processes, which is itself determined by their degree of subsumption or structuring by the dominant capitalist relations of production. The requirements of the capitalist mode of production for enlarged reproduction (reproduction on an ever expanding scale) necessitate a constant decrease in the time devoted to the reproduction of labour-power in the labour process, this being achieved by such means as increasing the productivity of labour. Increases in the latter create the possibility for an increase in the variety and range of commodities available for individual consumption, through a constant structuring of the labour processes to meet the requirements of capitalist production. Thus, individual consumption, which appears to connect the use-values produced with human needs in a fairly immediate way, is dependent upon the technical level achieved by production in the labour processes, which is itself a result of the degree to which the labour processes have been (or can be) transformed by the dominant relations of production (we will return to this point in greater detail later in the text). In much the same way that this individual consumption is governed by the process of production, so too is productive consumption. Here the consumption of use-values as means of production (which thereby ensures the reproduction of the conditions of production) is structured by production; the proportion between departments I (production of means of production) and II (production of articles for individual consumption) is determined by the nature and quality of the use-values required as means of production, which is again dependent on the technical capacities of production—upon the degree of transformation of the productive forces by the dominant relations of production—and the needs of personal consumption through availability, distribution and the modes of consumption promoted by the agents charged with marketing the commodities produced.

With regard to exchange, or the process of circulation of the product, it exists as an intermediate phase between production and distribution on the one hand, and consumption on the other, the latter being determined, as we have seen, by production itself. Whether it is a question of exchange of commodities between departments I and II, or between capital and labour power in the process of production, or of exchange between income and commodities for personal consumption, exchange always operates within limits set down by the process of production. Marx formulates this very concisely in the following passage, which is worth quoting in full:

> Firstly, it is evident that exchange of activities and skills, which takes place in production itself, is a direct and essential part of production. Secondly, the same applies to the exchange of products in so far as this exchange is a means to manufacture the finished product intended for immediate consumption. The action of exchange in this respect is comprised in the concept of production. Thirdly, what is known as exchange between dealer and dealer, both with respect to its organisation and as a productive activity, is entirely determined by production. Exchange appears to exist independently alongside production and detached from it only in the last stage, when the product is exchanged for immediate consumption. But (1) no exchange is possible without division of labour, whether this is naturally evolved or is already the result of a historical process; (2) private exchange presupposes private production; (3) the intensity of exchange, its extent and nature, are determined by the development and structure of production, e.g. exchange between town and country, exchange in the countryside, in the town, etc. All aspects of

> exchange to this extent appear either to be directly comprised in production, or else determined by it.[15]

From this example of the capitalist mode of production, we can see how a particular process of production can determine forms of distribution, consumption and exchange. In any mode of production, there is a specific combination of these forms, structured by production itself.

To add to this extremely schematic outline, we should note that in any mode of production, the determining element (production) can also be determined by the forms that it itself determines. For example, Marx shows in *Capital*, Volume 2, how the production period (or time of production) for a given commodity in the capitalist mode of production can be determined by the circulation process.[16] Yet, this process of circulation exists and always operates within limits that are determined by production. In the capitalist mode, for example, enlarged or simple reproduction of the production process necessitates (and thereby determines the existence of) a permanent maintenance of monetary capital and its circulation between the various units of production: as a result of this, the time and rhythm of this circulation, the expansion or contraction of monetary capital, can then exist as negative limits to the contraction or expansion of the time of production itself, which clearly must effect (amongst other things) the level of productivity, the time of turnover of industrial capital within this production process, etc. Thus, we see here that *the determinant* (production) *is itself also subject to determination, but within limits ultimately set by the determinant itself*. As Marx states:

> A distinct mode of production thus determines the specific mode of consumption, distribution, exchange and the specific relations of these different phases to one another. Production in the narrow sense, however, is in its turn also determined by the other aspects.[17]

The concept of a mode of production is not solely concerned with analysing the relations it requires between production, consumption, distribution and exchange. It also includes a theorisation of its history,[18] which, following Balibar, we can term its genealogy—and of its historical tendency, an effect which exists as a result of the reproduction of the structure of the mode of production which can be termed its dynamics.

The *genealogy* of a mode of production can be briefly defined here as consisting of the history of those *particular elements which combine to form a specific mode of production*. Its field is the history of the transition from the previously dominant to the contemporarily dominant mode of production within a given social formation. Its object is to analyse how the elements of the existing combination emerged from the process of dissolution of the previous mode of production. To take a brief example (since we will be concerned at length with the genealogy of the capitalist mode of production in a later chapter): one of the elements of the combination existing in the capitalist mode of production is the direct producer separated from ownership and control of the means of production he operates; the genealogy of this element would examine the history of the creation of this wage-labour, of the historical process of separation of labourers from their means of production, which is a prerequisite for the existence of the capitalist mode.

In contrast, the concept of the *dynamics* of a mode of production has as its object the specific effects of the operation of a definite mode of production upon the reproduction of that mode. We have seen that a mode of production is characterised by a specific combination of elements, in which particular relations of production dominate the labour processes and ensure a particular form of extraction and appropriation of surplus labour from within these processes. The subsumption of the latter under these dominant relations, and the reciprocal limitation of the relations of production by the process of real appropriation within the labour processes, gives rise—in the case of each mode of production—to a development tendency which is specific to it and differs from one mode to another. The operation of this tendency has definite effects on the historical development of the mode of production, setting limits on its expansion and generating economic contradictions that are repeatedly manifest throughout the period of the mode's dominance in a social formation. Both the tendency and its effects constitute the dynamics of a mode of production. We will develop this concept in greater detail later in the chapter when we analyse the dynamics of the capitalist mode of production, and the possibility for different forms of capitalist economic penetration of non-capitalist modes existing within it.

Having outlined schematically the concepts required to analyse modes of production as forms of combination of the elements required for the reproduction of economic practice, and indicated how these forms differentially structure the social formations in which they are dominant, we can now go on to examine the other practices of the social formation that are ultimately determined by the mode of production.

Ideological practice

Here we can be extremely schematic, since we are largely restating theses that have been formulated elsewhere.[19] Rather than conceiving ideology—as is generally the case—as a form of illusion or imaginary construction external to reality, we can define it as a practice that is relatively autonomous and has a material existence.

As with the economic, ideology is a practice with a specific object: to transform the subject's existing relations to the lived world into a new relation, through the process of ideological conflict. Its function in any social formation is to constitute individuals as subjects.

To give a brief example of this: we noted earlier that the real subsumption of the productive forces under the dominance of capitalist relations of production (i.e. under the dominance of the elements of economic practice that combine to produce a specifically capitalist mode of production) can, at a certain stage, establish a combination within the labour process between means of labour and object of labour, in the form of mechanised industry. The corollary of this, of course, is that the economic subject required for the reproduction of such a unity is quite different from that required by more simple forms of co-operation in the labour process, in which the worker possesses his own means, and brings them to this process. Whereas the latter would require a form of labour closer to artisan forms (the ability to set the means of production to work belonging to the individual), the former requires a notion of labour as human activity brought to a set of instruments and materials which exist apart from labour and determine the latter's conditions of work. It would seem, therefore, that the economic subject, the specific form of economic individuality required by the labour process for its reproduction is determined by the combination existing in the labour process at any one moment in its development.

In much the same way, it can be shown how other practices of the social formation, in their historical development, require forms of historical individuality that are specific to them.[20] This process, however, is made more complex by the fact that each of these practices intervenes within the other. At any moment, the constitution of historical subjects is a result of an inter-relation of different practices which are themselves structured by the determinant instance(s) in the social formation. This structuring by the determinant instance is, in turn, ultimately determined by the combination of elements specific to the mode of production.

In order that the social formation itself can be reproduced, it is essential that the concrete subjects that are supports for the various practices (and their inter-relations) interpret their own lived relations to the world as a function of their position, as defined by the requirements of the structure. It is the particular function of ideology to produce such an interpretation. The object of ideological practice in a specific social formation is to ensure that concrete subjects represent to themselves in ideology their conditions of existence. It is by doing this that ideology ensures the reproduction of the social formation. Thus, as with the other practices, the ideological has a structure and object (ideological transformation) that is specific to it—and, in this sense, is autonomous; Yet, this autonomy is relative, since the specific content of ideologies, the time, direction and limits of their transformation is, as with the other practices, subject to the structuring effect of the instance(s) that is (are) determinant in a particular social formation subject to the dominance of one mode of production.

If ideology exists as a practice, then it must have a material existence. Consequently, in any social formation, ideologies always exist in definite apparatuses.[21] The concrete subject lives within ideology—not within one ideology, but within a complex overdetermination of different ideologies—and, as a result, he adopts particular 'attitudes', and participates in events governed by material rituals (voting, housework, purchasing a commodity, etc.) which always exist within an apparatus (the family, the church, the trade union, the school, etc.). It is by participating in these practices and rituals that the subject is confirmed in the correctness of his lived relations to the world (within their respective ideologies, practice has proved for many years that parliament represents the wishes of the people, that men are rationally superior to women, production is geared towards consumer needs, and so on). Consequently, ideology always exists in the form of material actions, governed by practices and rituals that are inserted into apparatuses, which are always ultimately subject to the ideological requirements for the reproduction of the dominant mode of production, to the requirements of the dominant relations of production, and, hence, ultimately to the ideological forms within which the class(es) that controls the production and distribution of the surplus product lives its relation to the real world. In every social formation in which classes exist, it is always this latter overdetermination of ideologies that seeks to mould all other ideologies into an image of itself, through seeking to gain ideological hegemony within the apparatuses.

Political practice

Political practice similarly has a specific object—the occupation of state power through a transformation of the existing balance of social forces, in order to establish or perpetuate the dominance of a particular class or alliance of classes represented politically in the state apparatus. This balance is determined in the last instance by the determinant relations of production installed through the dominance of a particular mode of production. This, however, does not require a simple linear causality between economic and political practice, between something termed an 'economically dominant' class and a 'politically dominant' class. Such a conclusion would be inadequate, for a number of reasons. First, it is generally the case that political hegemony within the state apparatus is exercised by a class, or fraction[22] of a class that is not economically dominant. (Examples are legion: Gaullism, early German and Italian fascism, late Argentinian Peronism, nineteenth-century Bonapartism, early twentieth-century English liberalism, seventeenth-century absolutism, etc.) Secondly, beyond these 'exceptions', it is always the case that, even if a fraction of the economically dominant class (or, indeed the dominant class itself) is politically dominant, its dominance rests upon alliances with other classes which are represented at the political level. The dominant class (in its various fractions) attempts to occupy and transform the state apparatuses so that these can function to ensure the reproduction of the dominant relations of production; nevertheless, this dominant class can—and must (as we shall see later in the text when we examine the transition from dominance by a feudal to dominance by a capitalist mode of production)—perpetuate this objective in a political alliance with other classes, even under the political hegemony of another class (or fraction of a class) in the state apparatus.

Whilst political practice is relatively autonomous, its development, its interventions in other practices, is determined and structured by the determinant instance. As a result, limits are placed upon the political in its interventions in other paractices and these are determined by the particular instance that is required by the reproduction of the dominant mode of production to occupy the place of determinance in the social structure. For example, it can clearly be the case that, under the dominance of monopoly capitalism, the financial capitalist and industrial capitalist fractions of the capitalist class do not constantly exercise political hegemony; nevertheless, the political practice of whichever class of fraction occupies the hegemonic position is limited in its actions in relation to economic practice by a whole series of objective co-ordinates related to the organisation of economy and society and the state's role in this organisation. The state is not, therefore, a simple instrument that the politically hegemonic class or class alliance can utilise at will, since its intervention in the economic must take place within limits established by the stage reached by the capitalist mode of production at the time of its intervention. Similarly, the interventions of political practice within ideology, within the ideological apparatuses, are ultimately limited—as we have already noted—by the lived relation to the world constituted for subjects by the ideological requirements for the reproduction of the dominant mode of production in the particular time of its dynamics. Thus, as with the other practices, the interventions of the political are confined within limits ultimately set by the determinant instance in any particular social formation.

We stated at the beginning of this brief outline of the theoretical object of historical materialism, that the structure of every social formation contains four relatively autonomous practices. In this text I do not examine the structure of *theoretical* practice—that practice which, by working on raw material constituted in theoretical or philosophical ideologies transforms these into systems of scientific knowledge, which then enter into the reproduction of the other practices. Much has already been written on the relation of scientific to non-scientific philosophical and theoretical discourses on the one hand, and to ideological discourse on the other,[23] and, in any case, this issue is peripheral, given our objective here. In a subsequent text, which will try to use the theoretical framework developed here to analyse a particular Third World formation, the problem of the relation between theoretical analysis and its insertion into non-theoretical ideological discourse will necessarily have to be posed. Given that the object of the present text is, however, to outline this general framework, this problem is beyond its scope. Consequently, we are solely concerned with the other practices of the social formation.

THE HISTORICAL TIME OF THE SOCIAL FORMATION

This entails posing the following question: what is the *concept of historical time* peculiar to the structure of the social formation?—'concept'—since, as we have noted, there is no empirically given raw material that presents itself as 'history'. Historians necessarily operate with a particular ideologically constructed idea, notion, or concept of historical time that they bring to their data. We have seen that the different practices of the social formation are relatively autonomous.

The corollary of this, of course, is that they cannot all simply be thought in the same historical time, and that, consequently, each practice has its own relatively autonomous history and differential historical existence.

Here, however, we are approaching a difficult problem, since, although it is clear in a descriptive manner that each of these relatively autonomous practices has its own specific history, this can only really be known rigorously if we have constructed the concept of the historical temporality of each of these practices under the dominance of a particular mode of production. Yet, with the exception of the analysis of the economic instance in the capitalist mode, and the analyses by several theorists of particular discourses,[24] this work remains unachieved. In the absence of established concepts of the different times of the practices of the social formation, it becomes impossible to theorise rigorously the specific forms of intervention of these practices within each other in the history of any one social formation. All we can do—as we have briefly indicated earlier—is to show how, given the dominance of one mode of production, limits are placed on the forms of intervention of the various practices within each other, limits which are ultimately determined by the requirements for the enlarged reproduction of this mode, which (as we have seen) is determined by the structure of the economic instance itself, since it is the latter that determines which of the practices will occupy the determinant place. Thus we can say that ultimately, both the structure and the historical times of the relatively autonomous practices are structured by the articulation of the determinant instance within the social formation. As a result, there is a definite form of dependence established by the articulation of the determinant practice within the structure which always exists in the historical time of this structure.

This conception is of fundamental importance if we are to approach the problem of the uneven development of the various practices of the structure in relation to each other. Given what we have said above about the relative autonomy of the different practices, it is clearly the case that, in terms of their historical times, any one practice (or one of the forms within which this practice exists) can be dislocated in relation to another. For example, the practico-social and theoretical ideologies that constitute ideological subjects who will interpret their lived relations to the world in terms of the ideological requirements for the reproduction of the labour process dominant at a particular time in the development of the capitalist mode of production, may not be adequately secured within the ideological apparatuses, in that their reproduction may remain partially blocked by the dominance of pre-existing ideologies; again, the class alliance represented as dominant at the political level may hinder the process of subsumption of the productive forces under the dominant relations of production. Similarly, transformations can take place in the legal ownership of the means of production without this necessarily transforming the relations of production under which the surplus product is extracted, and so on. Yet, all such dislocations, and the limits within which they can operate, are ultimately governed by specific preconditions. In every social formation, it is *the practice which occupies the determinant place that ultimately structures this process of dislocation*, despite the fact that the uneven development the latter engenders can produce the dominance of a particular instance at a specific moment in the historical time of the social formation. It is only by analysing the basis of this dominance in its determinancy by the determinant instance, that we can show how one concrete situation (or conjuncture) in the development of the social formation is distinguishable from another concrete situation (in which a different practice or one of the forms of this practice is dominant) and how the relationship between the dominant instance and the other instances in this concrete situation can only vary within limits established by the determinant instance. This, however, is to go beyond our objective in this section.

All we have tried to establish schematically here is the structure of the social formation itself. We can conclude very briefly by stating that the social formation is a structure containing a number of practices which have variant structures and are relatively autonomous, being determined in the last instance by economic practice—or, rather, by the particular combination within which this practice exists as a mode of production. Every social formation has a historical time which is characterised by times specific to different practices, and the intervention of these practices within each other. However, both these times and interventions are subject, in their variation, to limits imposed by the articulation of the determinant instance within the social formation. The instance which occupies the determinant place is itself determined by the structure of the mode of production that is dominant in the social formation.

SECTION II: THE DYNAMICS OF THE MODE OF PRODUCTION: THEORISING FORMS OF CAPITALIST PENETRATION

In the social formations of the Third World, there appears to be a structuring not by one mode of production, but, as we indicated in

Chapter 4, by a combination of different modes of production. This combination has been produced by different forms of capitalist penetration within what were previously non-capitalist formations.

Our object here is to analyse the effects of these different forms in producing the combination of modes of production, to examine how one analyses this combination, and the way in which it structures the other instances of the social formation.

We are not, therefore, primarily concerned with analysing the determinants of capitalist penetration in its various forms. Nevertheless, we do need to outline a theoretical framework in which we can at least situate the basis for the tendential development of the capitalist mode of production producing a restricted and uneven capitalist development in a social formation previously structured by a non-capitalist mode of production. In an earlier chapter, we have indicated the means by which industrial capitalist economies restrict the development of Third World economies. Now, having outlined the concept of mode of production, we need to show how the basis for this 'restricting' can be theorised.

Consequently, the remainder of this chapter will outline how the possibility for capitalist penetration of non-capitalist modes *can be theorised from within an analysis of the dynamics of the capitalist mode of production*, indicating how this penetration has taken varying forms, as determined by different stages in the development of these dynamics. Having done this, we can then go on to examine the effects of these forms in producing the combination of modes of production that structures Third World formations.

THE DYNAMICS OF THE CAPITALIST MODE

In order to analyse the possibility for the major forms of capitalist penetration of non-capitalist formations, it is essential to examine the dynamics of the capitalist mode itself. What follows is a brief resumé of Marx's analysis in *Capital*, which provides the basis for our investigation.

In Volume I of *Capital*, Marx analyses how the productive forces are increasingly subsumed under capitalist relations of production in order to meet the requirements of enlarged reproduction for a perpetual increase in the productivity of labour. In Volume 3, he goes on to show how this development is expressed in an increasingly higher organic composition of capital (in a relatively greater increase of constant to variable capital), the immediate effect of which is a tendency for the rate of profit to fall in all the major spheres of production. Thus, accompanying the accumulation of capital, the socialisation of the productive forces, and the extension of capitalist relations to all branches of production, there is a concomitant tendency for the rate of profit to fall within the capitalist mode.

The tendency follows, in Marx phrase, 'as a logical necessity' from the structural development of this mode of production, for the following reasons: in the capitalist process of production, the capitalist, through his ownership and control of the means of production, sets living labour to work on a mass of materialised labour (constant capital—both fixed and circulating). The rise in the organic composition of capital attendant upon the increase in the social productivity of labour produced by this application leads, however, to a decline in the quantity of the living to the materialised labour used up in the process of production. If we assume that the average working day, average level of wages and the proportion between necessary and surplus labour (i.e. the rate of surplus value) all remain constant, then it follows that, although the mass of surplus value realised from living labour will increase as a result of the increase in materialised labour, the rate of profit will, in fact, fall, since it is only living labour, and not materialised labour that is creative of value. (The same effect will, of course, be produced to a lesser extent with an increase in the rate of surplus value, as long as this is relatively less than the concomitant increase in the organic composition of capital.)

Marx, however, speaks of a 'tendency', since its development is weakened, or 'checked' by what he terms 'counterbalancing forces': an increase in the intensity of exploitation (through a prolongation of the working day, an increase in relative surplus value by improvements in production methods without any alteration in the magnitude of the capital invested, etc.); the depression of wages below the value of labour-power; the cheapening of the elements of constant capital; the utilisation of labour-power displaced by an increase in constant capital being used to open up new sectors of production—which Marx terms 'relative over-population'; the increase of stock-capital, and foreign trade.

It seems, therefore, that the development of the structure of the capitalist mode of production, both produces this tendency and the counter-effects to its operation. For example, a rise in the organic composition of capital is accompanied by an increase in relative surplus value and (possibly) an increase in relative over-population which, enabling new sectors of production to be opened, counteracts a falling rate of profit. Similarly, existing capital can be depreciated as a result of an increase in relative surplus value; this depreciation can then cheapen the various elements of constant capital, which will prevent the value of constant capital from increasing in the same proportions as its material volume, and so on. The crucial question now becomes the importance of this tendency for the dynamics of the capitalist mode of production. The operation of the tendency and its counter-effects produces what Marx terms 'rhythms' of production, in which different combinations of counter-effects can operate against the tendency at different times. There always exists a specific 'equilibrium', as a particular state of combination of these counter-effects.

Having schematically outlined the dynamics of the capitalist mode, we can now go on to examine how the possibility for the economic existence of different forms of capitalist penetration of non-capitalist modes must initially be situated within this dynamics. At any given moment in the dynamics of the capitalist mode of production, there exists a particular combination of counteracting effects; the possibility for capitalist penetration and the varying forms this can take, must initially be analysed from within this combination. We can illustrate this by examining how the possibility for one form of capitalist penetration—namely imperialism—can be analysed in this way. As raw material, we can take Lenin's analysis in *Imperialism* . . . ,[25] and Hilferding's work in *Finanzkapital*.[26]

IMPERIALISM AS A SPECIFIC FORM OF CAPITALIST PENETRATION

As is well known, Lenin defines imperialism as a stage in the development of the capitalist mode of production, which is characterised by five basic features: (1) The concentration of production and capital in the form of monopolies; (2) The merger of banking and industrial capital to form finance capital; (3) The dominance of capital export over commodity export; (4) The formation of industrial monopolist associations; (5) The territorial division of the world amongst the most advanced capitalist social formations. This characterisation has its own theoretical prerequisites, which, for the purposes of our analysis, we must uncover.

The concentration of capital and its determinants

Firstly, take the concentration of industrial capital. What are its preconditions? On the most fundamental level, Marx gives us the answer to this in Volume 3 of *Capital*, where he shows how a fall in the rate of profit is accompanied by an increase in accumulation and a higher organic composition of capital which provide a basis for increasing productivity and a rise in the mass of profit. This process

> . . . requires a simultaneous concentration of capital, since the conditions of production then demand an employment of capital on a larger scale.[27]

The important point for us is that this process of concentration (accompanied as it is by a process of centralisation of capitals in which small capitals with higher rates of profit are subsumed under larger capitals with higher rates of accumulation but lower rates of profit) establishes improved conditions for the operation of the counter-effects or, more specifically, it extends the limits within which they can operate. This is clear from Marx's arguments in *Capital*, and from Hilferding's analysis in *Finanzkapital*. The costs of the elements of constant capital and its application can, for example, be decreased by economies in the use of constant capital (expenditures on raw materials, operation of fewer and larger elements of capital equipment, savings on buildings, storage, etc.). They can also be decreased by the reduction in circulation and depreciation costs made possible by concentration and centralisation, and by the reduction in the overall costs of capital equipment resulting from increases in productivity due to concentration in other sectors. Again, the concentration of capital, through its relative increase of constant to variable capital, promotes the release of labour-power from the relatively higher capitalised sectors, thereby enabling new sectors of production to be opened up with initially higher rates of profit (which are then themselves potentially susceptible to capitalisation, etc.). Similarly, capital concentration and centralisation, by establishing the basis for the economic form of monopoly (although concentration and centralisation can, of course, be represented in other economic forms), can intensify foreign trade and capital investment within other social formations, thereby lessening the tendency for the rate of profit to fall. At the level of the monopoly form, this can also result in effective measures to control wages, and so on.

This concentration and centralisation of industrial capital which extends the operational limits of the counter-effects can, however, only exist on the basis of what Marx terms a particular 'social combination of labour'. We argued earlier that a mode of production exists as a combination of labour processes and a dominant mode of extraction of surplus labour, and that within this combination, the mode of extraction 'structures' a combination of labour processes, transforming them within limits set by the requirements of the dominant mode of extraction. In terms of our earlier definition of the different forms of labour process possible under the dominance of capitalist relations of production,[28] we can trace a development from a simple to a more complex co-operation in the division of labour. Despite the differences between them, however, both forms unify labour-power and the means of labour; the instruments of production are operated by the workers, and their effective use ultimately depends on the workers' knowledge and skill. At a certain stage, however, a displacement occurs within this combination of elements founding both simple and complex co-operation. The unity of labour-power and means of production is replaced by a new unity—that between the means of production and object of labour, which is produced by mechanisation. This new combination, enabling the application of specific discoveries to means of production in the form of technology, is the social combination which Marx refers to as establishing the basis for capital concentration and centralisation. Elsewhere he terms it 'socialised labour', the result of a long process of transformation of the labour processes by the dominant capitalist relations of production. Thus, the process of concentration of industrial capital, which, as we stated earlier, extends the limits of operation of the effects counteracting the tendency for the rate of profit to fall, appears to have a basic precondition, namely *the existence of a particular combination of the elements of the labour process*. Furthermore, it seems that this combination can only exist within a relatively more 'advanced' capitalist mode, when the labour processes have become fully subsumed under capitalist relations of production.

The unity of industrial and banking capital

The possibility of the existence of imperialist penetration does not only depend on the concentration of industrial capital, but also requires what Lenin calls the 'merger' of industrial and banking capital or, as Hilferding terms it, 'the unity of industrial and banking capital'. The preconditions for this 'merging' must, therefore, be specified.

The functions fulfilled by banking capital in the processes of circulation and production are elaborated by Marx in Volume 3 of *Capital*, and there is no need to repeat them in any detail here. The 'savings' of those classes who do not control the process of capitalist production are made available for accumulation and investment; the costs of capital circulation are reduced; interest-bearing and fictitious capital are created through the selling of stocks, shares, bonds and the provision of loans (these then providing the basis for further capital investment); and the barriers to the equalisation of profit rates which would otherwise exist as a result of the immobility of industrial capital are raised.

Alongside these basic functions, with the development of capitalist production, the need for increased accumulation constantly requires a large banking capital which can supply industrial capital with the credits needed for payment and production. This creates the basis for a concentration of banking capital, alongside the concentration of industrial capital. These two structurally inter-related processes of concentration establish the basis for the 'merging' that Lenin discusses in *Imperialism* . . .

Hilferding draws out the results of this very succinctly:

> A steadily increasing proportion of capital does not belong to the industrialists who use it. They can dispose of the capital only through the bank, which represents the owners of the money to them. On the other hand, the bank must place a steadily increasing part of its capital in industry. In this way, the bank itself is growing, to an ever increasing extent, into an industrial capitalist itself. I call this *bank capital, that is capital in the form of money which in this way is in fact transformed into industrial capital, finance capital*. . . . A steadily growing part of the money used by industry is finance capital, capital at the disposal of the banks and used by the industrialists. . . . Industrial Profit gains a more secure and stable character, and the investment opportunities open to bank capital in industry also expand steadily. . . . It is thus clear that, with growing concentration of property, the owners of the fictitious capital which gives control over the banks and of the capital which gives power to industry, become more and more identical.[29]

Once this unity has been produced, it has important results for the operation of the counter-effects to the rate of profit to fall, in that it greatly facilitates the movement of capital from one sector to another. This is of crucial importance for further capitalisation which is, of course, a prerequisite for any increase in relative surplus value—a major counter-effect.

Thus far, we have argued that the concentration of industrial capital, which itself extends the limits within which the counteracting effects can operate, can only develop given a combination of the labour process that exists when the productive forces have been fully subsumed under capitalist relations of production. These preconditions also govern the concentration of banking capital. The two concomitant processes of concentration, through their inter-relations in the processes of production and circulation, can be unified in the form of finance capital, but this result can only emerge at a specific time in the dynamics.

Capital Export

Having reached this point, we can now go a step further, and show how the existence of finance capital establishes the possibility for a relatively greater degree of capital export than existed prior to its emergence within the capitalist mode of production.

Concentration, as we have seen, greatly increases the rate of accumulation. The accumulations of finance capital resulting from the increase of accumulation attendant on concentration can be utilised internally, although there are limits to this. There is an ultimate limit point, when

> Capital [has] grown in such a relation to the labouring population that neither the absolute working-time supplied by this population, nor the relative surplus working-time, could be expended any further; at a point, therefore, when the increased capital produced just as much, or even less, surplus-value than it did before its increase, there would be absolute over-production of capital.[30]

Such a point marks the end of a production-cycle, capital being depreciated and production recommencing at a lower level, etc. This ultimate limit, however, need never be reached for unused accumulations of finance capital to be exported, since there are other structural limitations placed upon its internal use. For example: the extent to which capitalist relations of production have penetrated and transformed the agricultural sector, the degree to which existing sectors are conducive to capitalisation,[31] and so on.

These limits, as economic barriers to internal utilisation, cannot, of course, in themselves, provide a satisfactory basis for explaining why accumulations of finance capital are exported at specific times within the imperialist period. For example, it was clearly not the case during late nineteenth-century European imperialism that capital was exported because these structural limits had been reached. To examine why, at a particular economic moment, capital is exported, we need to introduce additional elements into our argument.

In Volume 3 of *Capital*, Marx states,

> If capital is sent abroad, this is not done because it absolutely could not be applied at home, but because it can be employed at a higher rate of profit in a foreign country.[32]

Hilferding and Lenin, arguing against the underconsumptionist rationale for capitalist penetration, also cite this as the basic determinant of capital export. Once its possibility is established, its realisation requires either the existence of higher national profit rates outside its own formation, or, at the very least, potentially higher rates in the long term. This situation clearly existed for capital export from industrial capitalist modes of production in late nineteenth-century Europe. The free trade era of commodity export had virtually been terminated by the period 1860–70, as a result of the industrialising countries of Europe and the U.S. erecting tariff barriers against imported commodities to protect their internal markets. This process, which effectively subsidised indigenous production by enabling European monopolies to sell their commodities above their values and not be undercut by foreign imports, led to increased accumulations of finance capital for export in Europe and the U.S. This capital then attempted to penetrate other less advanced industrialising modes of production which, with their lower organic compositions, could yield higher rates of profit on the capital invested. Yet, limitations were, again, put on this penetration by the less advanced capitalist social formations, which acted to block this expropriation of surplus value produced in their own modes of production.

Consequently, capital export turned to the non-capitalist modes of production that—to a greater or lesser extent—had been drawn into the world economy and division of labour set up during the period preceding the dominance of finance capital. Here the potential for realising increased rates of profit was vast. The non-capitalist social formations of Latin America, Africa and Asia were characterised by a potential availability of labour-power at cheaper rates than those in capitalist formations, and the low quality of this labour was offset by extremely long working hours, low ground-rents, a high availability of land, and relatively cheap raw materials. Once captured, a non-capitalist formation could then be protected, thereby ensuring that its raw materials could be exploited solely for the enlarged reproductive needs of the 'home country', and its consumption needs met solely by goods either produced in the latter, or by new units of production set up by finance capital invested in the non-capitalist mode of production.

Yet, whilst the potentialities for investment were immense, so were the barriers to it. Capital could not be invested to meet these objectives if capitalist forms of production did not exist and, as we shall see, as a result of the articulation of the dynamics of many of these non-capitalist modes within their respective social formations, the basic prerequisites for the existence of capitalist production were non-existent. Imperialist penetration, faced with this situation thus set about the task of forcibly creating these prerequisites. This is what clearly distinguishes it from other earlier forms of capitalist penetration of non-capitalist formations. This, however, is to anticipate what follows.

Our main object here has been to show how the possibility for the export of capital and imperialist penetration can be analysed as a necessary effect of the capitalist mode at a particular stage in its dynamics, which we have attempted to characterise. All we have tried to do is to indicate how the material basis for one particular form of capitalist penetration can be located in the dynamics of the capitalist mode. Beyond this foundation, the specific forms taken by this penetration can only be examined through a conjunctural analysis, not only of the social formation from which capital is exported, but of the relations existing between it and the relative (economic, political and ideological) development of its capitalist competitor nations, and that of the non-capitalist modes of production and social formations that it attempted to penetrate. Such an analysis, having established the basis for the possibility for the export of large amounts of accumulated finance capital would then have to deal with the particular economic forms in which this process was represented (viz. monopolies, monopoly forms subject to state intervention, the 'multi-national corporation', etc), the complexity of the relationships between these different forms and the state, and the particular combination of ideologies within which this export, its forms, and the resultant forms of penetration under its dominance were represented (a problem that Lenin initially approaches in his concept of the 'labour aristocracy'[33]). Similarly, the extent to which the export of capital to less advanced capitalist or non-capitalist modes of production could successfully penetrate can only be fully analysed if we undertake similar conjunctural analyses of their concrete situation at the moment of imperialist penetration. These analyses must initially be based on the articulation of the dynamics of their respective modes of production, since this established the possibility (or otherwise) of creating the prerequisites for capitalist production.

All these problems, concerning the conjunctural determinants of imperialist penetration, which would have to be examined in any analysis of the late nineteenth-century emergence of imperialism, are not part of the object of this chapter. All we have tried to do here is to establish a theoretical basis for analysing why the possibility for capital export can only emerge at a particular stage: when certain preconditions are established within the dynamics of the capitalist mode of production.

Our conclusion is that imperialism—as one form of capitalist penetration of non-capitalist modes which, as distinct from other, earlier forms of penetration, is characterised predominantly by the export of capital—can only exist and produce its specific effects at a particular time in the dynamics of the capitalist mode, a time defined by a specific combination of the operation of the counteracting tendencies, a combination that is itself determined by the degree of subsumption of the productive forces. Our original contention, that the possibility for the existence of a particular form of capitalist penetration must initially be analysed as resulting from a particular combination of the operation of the counter-effects within the dynamics of the capitalist mode, seems, at least, to be valid for one of these forms—that of imperialism.

This conclusion should also apply to other forms of capitalist penetration of non-capitalist modes that have occurred in the genealogy and dynamics of the capitalist mode.

COMMODITY EXPORT AS THE DOMINANT FORM

Take, for example, the form of capitalist penetration of non-capitalist modes that preceded imperialism. As is well known, the dominant place was held here by the export of commodities rather than the export of capital, the former being realised through an exchange of non-equivalent values with other modes of production. The conditions under which production took place in the relatively more advanced capitalist mode of production enabled the latter to export commodities to other less advanced capitalist, or non-capitalist formations where they could be sold at a price which, although above their value, would be below the price and value of equivalent products produced in the less advanced capitalist or non-capitalist mode. Meanwhile, whilst the latter modes exported commodities to the more advanced capitalist mode of production, they nevertheless offered more materialised labour in the form of commodities than they received, due to the higher level of productivity in the industrialised capitalist mode. As a result, the latter received the raw material elements of its constant capital at a cheaper price, and could also reduce the value of the variable capital employed in production, since the agricultural commodities imported lowered the value of labour-power. Both these results tended to raise the rate of profit in the short term, just as the export of capital does in the imperialist period. Capitalist penetration of non-capitalist formations took this particular form during the pre-imperialist period (or what is more popularly known as the 'competitive capitalist' stage[34]) of the capitalist mode of production. In the case of British industrialisation, we can set general limits to this period, from 1780–1850.[35] Whilst the unequal exchange of values represented in the commodity form persisted after this date, with the growing importance of capital export it became increasingly subsumed under the requirements of the latter.

In much the same way that we theorised the economic basis for the dominance of capital export, it should similarly be possible to theorise why, how, and to what extent at different times during this period, the effect of the export of commodities was produced as the dominant form as a result of the operation of the counter-effects within the dynamics of the capitalist mode, in relation to a particular level of subsumption of the productive forces. Marx speaks of 'the innate necessity for an ever expanding market', but he never specifies why this 'necessity' exists and how it can vary temporally. What is required, therefore, is an analysis of its foundation within the dynamics of the capitalist mode, in much the same way that we did for the concept of capital export during a later period. Although this cannot be done within the limits of the present text, our analysis of the imperialist period shows this theorisation is possible from within the dynamics of the capitalist mode.

THE DOMINANCE OF MERCHANTS' CAPITAL

Similarly, those forms of penetration that occured in the time of the *genealogy* of the capitalist mode, under the dominance of what Marx terms merchants' capital, can also be theorised.

As is well known, this form of capital—with its payment of money(m) for a commodity(c), which is to be sold at a price above its initial cost (m – c – m)—can exist in many different modes of production, since its only prerequisites are the circulation of commodities and the existence of money.[36] What particularly interests us here is the role played by merchants' capital in what we have termed the genealogy of the capitalist mode of production. This genealogy—whose historical field within European social formations was the transition from feudalism to capitalism—is characterised by the development of the prerequisites for capitalist production, namely a separation of direct producers from their means of production, co-existing with increasing accumulation of capital. Marx terms this latter one a process of primitive accumulation. Thus:

> . . . it is evident, and born out by closer analysis of the historical epoch which we are now discussing, that the *age of dissolution* of the earlier modes of production and relations of the worker to the objective consideration of labour is *simultaneously an age* in which *monetary wealth* had already developed to a certain extent, and also one in which it is rapidly growing and expanding, by means of the circumstances which accelerate this dissolution.[37]

In the European transition to capitalism, this accumulation came from diverse sources—usury, national debt-financing, taxation—but, above all else, from commerce, since the sixteenth and seventeenth centuries in Europe were marked by a rapid expansion of commerce and trading on an increasingly world scale. This produced a rapid increase in the rate of mercantile capital accumulation, and developed a world market which would later be utilised by capitalist production. Occurring as it did alongside the process of separation of direct producers, this source of accumulation was crucial in the development of capitalist production. Later in the text, we will examine the forms taken by merchants' capital, and its effects on non-capitalist formations. Our main concern here, however, is to show how its dominance during this period can only be rigorously explained if it is analysed from within the genealogy of the capitalist mode of production, as it occurred in a particular transition from dominance by a feudal to a capitalist mode of production.

For a number of reasons, the accumulation and development of merchants' capital was initially restricted both by the feudal economic and social structure, and by the slow progress of capitalist production in many European countries during the sixteenth to eighteenth centuries.[38] Consequently, merchants' capital was generally forced to operate beyond the confines of its nation-state, and, as a result, it was during this period of dissolution and transition that trade under its dominance was at its peak. Thus, in the period of the genealogy of the capitalist mode in

Europe, merchants' capital was the characteristic form under which penetration of non-capitalist social formations occurred; it found its ultimate embodiment in the activities of such institutions as the English and Dutch East India Companies, English colonisation of the West Indies and the Spanish and Portuguese penetration of Latin America. Yet, reaching its peak during this period, merchants' capital and the control it exercised over commerce gradually became subjected to the increasing dominance of capitalist production. This led to its decline for a number of reasons.

First, the strength of merchants' capital lay in its control of its rate of profit, through monopolising selling prices and controlling costs of production, or the price at which the commodity was purchased. As capitalist production developed, however, the average rate of commercial profit became increasingly determined by the average rate of industrial profit (itself determined by competition between different industrial capitals[39]). Similarly, the costs of production were increasingly subordinated to the requirements of capitalist production. The decreasing influence of merchants' capital over the limits to its profit and the price of the commodity it purchased led to a colossal reduction in mercantile accumulation.

Secondly, industrial capital also attempted to reduce the activities of merchants' capital in the transporting, trading and marketing of the product, since any reduction in circulation costs increased the mass and even the rate of industrial profit. In this way, industrial capital attempted to reduce the amount of surplus value accruing to the merchant, and eventually to transfer the mercantile profit realised in the sale of the commodity into industrial capitalist profit. Consequently, merchants' capital, which was so crucial in the constitution and early development of the capitalist mode of production, gradually became subsumed under capitalist production, and the effects of penetration under its dominance were (as we shall see later) displaced by the effects of the rapidly developing export of commodities under the increasing dominance of what we have loosely termed 'competitive capitalism'.

From this brief statement, we can again see that the possibility for the existence of penetration under the dominance of merchants' capital must be situated within the temporal existence of the structure of the capitalist mode of production—not within its dynamics, but within its genealogy—within the process of constitution of the elements of the capitalist mode, as manifest in the historical forms taken by this process in the European transition to capitalism.

THE ECONOMIC EFFECTS OF CAPITALIST PENETRATION IN ITS DIFFERENT FORMS

Our analysis of the possibility for the existence of imperialist penetration tried to show how such penetration could only occur as a result of the operation of the counter-effects to the tendency for the rate of profit to fall within the dynamics of a capitalist mode of production in which the productive forces have been subsumed under capitalist relations of production. It also seems, at a general level that the forms of penetration of non-capitalist formations under the dominance of 'competitive capitalism' and 'merchants' capital' can similarly be explained from within the temporal existence of the capitalist mode, from within its dynamics at a less advanced time in the case of the former, and from within its genealogy in the case of the latter. These tentative suggestions on pre-imperialist forms of penetration cannot be developed any further here, but it is hoped that our analysis of the basis for one form indicates a direction in which further research can be undertaken.

There is, of course, a crucial element missing from the analysis so far. We have mentioned that the forms of capitalist penetration we have examined each have different effects upon non-capitalist modes of production. We will be analysing these effects at length in later chapters, but before concluding this chapter, it would be helpful to state in the most general way possible, prior to our theorisation of this problem, what we consider to be the major economic effects that result from these three forms of penetration of non-capitalist modes.

In general, penetration under the dominance of merchants' capital has as its major economic effect a reinforcement of already existing forms of extra-economic coercion, either through the introduction of types of landed proprietorship that appear to resemble late European feudal relations of production, or through a strengthening of existing relations of production in the penetrated non-capitalist social formation.

Penetration under the dominance of commodity export has as its general economic effect the gradual destruction of the existing circulation of commodities between the agricultural, rural artisan and urban artisan sectors of the non-capitalist mode. The object of this destruction is twofold: to transform the indigenous structure of production in order to promote the production of commodities for export to capitalist (and other) social formations, and to create a market for capitalistically produced commodities. In cases where, during this period, capitalist penetration meets a system of production in which the circulation and exchange of commodities is not prevalent—in, for example, a subsistence economy—it still attempts to transform this in a limited way, by encouraging the production of exchange-values for export. The extent to which a limited transformation of production by capitalist penetration can occur during this period ultimately depends on what Marx termed the non-capitalist modes 'solidity and internal structure', on its process of reproduction and the dynamics peculiar to it.

In general, penetration under the dominance of imperialism (the dominance of capital export) has as its overall economic effect the separation of direct producers from their means of production, and the foundation of an economic basis for a transition towards dominance by a capitalist mode of production.

Following on from the discussion of the structure of the social formation, we have briefly indicated how the concept of the dynamics of the capitalist mode can be used to specify the basic economic preconditions for two major forms of capitalist penetration of non-capitalist modes. In one sense, this was a digression from our main theoretical objective; the investigation of the combination of modes of production structuring Third World formations can take these forms as given, since it need only be concerned with their effects in producing the combination. On the other hand, since these effects are important for analysing the emergence of the combination, it seemed useful to put forward a framework—however tentative—in which the problem of the basis for their existence at different moments in the development of the capitalist mode of production could be posed.

Having outlined this in the briefest way possible, we can now return to our main line of analysis.

6 Transitional Social Formations Structured by an Articulation of Modes of Production

THE DISPLACEMENT OF THE DETERMINANT INSTANCE

When we begin to examine Third World formations, we are faced with a transitional social formation in which, rather than there being one dominant mode of production, there is a combination of different modes, capitalist and non-capitalist, in which the former is dominant or is becoming dominant. This transitional period is one in which a combination of modes of production is determinant in the last instance, and it is this co-existence, combination or *articulation*[1] of different modes within each other that provides us with a basis for thinking the structure and history of these social formations.

At the most general level, this transitional period, in which the formation is structured by an articulation of different modes, is characterised by what we shall term here a *displacement of the determinant instance*. This displacement, which results from the effects on the structure of the increasing dominance of one of the modes of production in the combination, is a basic characteristic of the transitional period. Consequently, we need to be clear as to exactly what we mean by this concept.

Within a given social formation, as we have seen, the structure of the dominant mode of production determines which of the instances occupies the determinant place; it is the structuring of the other instances of the social formation by the determinant instance that enables the mode of production to be reproduced. Now, assume that, as a result of the conjunction of a series of economic and non-economic processes—which may be largely internal to the social formation (as, for example, in the W. European transition from feudalism to capitalism), or external to it (as in the transition to capitalism promoted by imperialist penetration of Third World formations) a different mode of production gradually becomes articulated with the dominant mode of production, and begins to establish dominance over the latter. This new mode of production, in order that it can reproduce itself, will require, by its very structure, *determinacy by an instance other than the one that occupies the determinant place in this formation.* Thus, no matter what the processes are that promote the articulation of a new mode of production, ensuring its dominance in the social formation, the resultant transitional period must be characterised by a transformation of determinancy from one instance to another, if the new mode of production is to reproduce itself successfully. This process is a basic characteristic of transitional periods, around which analyses of particular transitions must be centred. In order to make this point a little clearer, we can briefly examine a particular example—that of the transition from feudalism to capitalism (we will be analysing this in detail later when we look at the genealogy of the capitalist mode).

Transitional formations: an undeveloped theoretical field

We have seen earlier that the reproduction of the feudal mode of production requires that the political instance occupies the determinant place in the social formation, since it is this determinancy that enables surplus labour to be extracted in the form of a ground rent levied on the direct producers. Conversely, we know that the capitalist mode of production requires a determinancy by the economic instance. Hence the transition from dominance by a feudal to dominance by a capitalist mode of production must necessarily effect a displacement of the determinant instance from the political to the economic, if the enlarged reproduction of a dominant capitalist mode is to be ensured. Once we approach this problem, however, we are effectively entering a new theoretical area, since, whilst historical materialist analyses provide examples of the effects of the subjection of the determinate relation of the mode of production to the structure and reproduction of another mode that is becoming dominant within a social formation, they nowhere systematically approach the problem of the transition through which this subjection or displacement occurs.

Marx, for example, indicates in *Capital*[2] that the determinate relation of production of the feudal mode continues to exist as a relation of distribution under the capitalist mode, but he never systematically examines how this transformation occurs. He shows how the previously dominant feudal landowning class undergoes a change in function, from controlling the process of extraction of surplus labour, to simply having a legal claim to a portion of the surplus labour that is produced and distributed, but he never approaches the problem of the transition that produces this change of function. Yet, in those sections of *Capital* (and in the *Grundrisse*), where he is not concerned with this problem, but with the genealogy of the capitalist mode—with the process by which the elements of this mode were initially constituted within the feudal mode, and the attempts that were made to secure the reproduction of these elements through the intervention of the political within the other instances[3]—he does provide a series of indicators as to particular areas to be analysed in approaching the problem of transitional social formations. We will be examining these in greater detail when we look at the genealogy of the capitalist mode of production.

The Effects of the Displacement

For the moment, however, we can take the main conclusion from Marx's analysis of the tranisition as given, and examine its relevance for our theoretical objective. It appears that in Third World formations penetrated by imperialism, with their structuring by a combined articulation of a non-capitalist and a capitalist mode in a transition in which the latter is generally becoming dominant, we must similarly have a tendency towards a displacement of the determinant instance.

As we shall see, the social effects produced by this displacement are complex. The instances of the social formation are characterised by a series of dislocations, in which, for example, at the economic level, the process of real appropriation no longer corresponds to the previously dominant relations of production. Similarly, at the ideological level, the overdetermination of ideologies that are dominant in the apparatuses may not constitute supports for the reproduction of the embryonic capitalist relations of production. Again, the intervention of the political instance within the economic may not be able to establish the preconditions for the reproduction of capitalist relations of production due to the contradictory class interests that are represented there, and so on. The existence of these dislocations, together with the fact that there is no inherent necessity about the transition—there is no reason why the combined articulation of capitalist and non-capitalist modes has to result in the dominance of the former—means that this combination can produce the dominance of a mode other than the capitalist mode: this possibility must be taken into account in any analysis of the future possible directions of change of the social formation. Having briefly characterised the social formation as being structured by an articulation of capitalist and non-capitalist modes of production which requires a displacement that produces a series of dislocations between the instances, we can now go on to the problem of the origins of this combination.

To answer this, we must turn to the historical field in which it was constituted, to the historical emergence of this combination. This field is formed by the *effects of the dynamics of the dominant mode of production in its articulation within the non-capitalist social formation.* Since this articulation is also effected by different forms of capitalist penetration, this must similarly be included within this field.

Before approaching this problem, however, we need to specify the general preconditions for the existence of a capitalist mode of production. This is crucial, since we can only begin to determine how the intervention of the various forms of capitalist penetration within the dynamics of non-capitalist modes can lead to the dissolution of these modes, and produce the particular form of combined capitalist-non-capitalist articulation that has come to dominate Third World formations, when we have outlined the pre-conditions for the emergence of the capitalist mode itself.

7 The Genealogy of the Elements of the Capitalist Mode of Production Within Non-Capitalist Social Formations

The preconditions for capitalist production

In an earlier chapter, we briefly outlined the structure of the capitalist mode of production, showing how it existed as a particular combination of the invariant elements common to all modes of production. We noted that the basis for this is a separation of the direct producer both from the object he produces and from the means of production he operates; the latter are owned and possessed by non-productive labour (even if this ownership and possession is exercised through agencies and functionaries) which, as a result of this control, is able to extract surplus labour in the form of surplus value from the direct producers.

Consequently—as is well known—in order that this mode of production can exist and be reproduced, two axiomatic 'preconditions' must be met:

(a) Monetary capital must be accumulated in the hands of non-productive labour which has effective control over the use of the means of production.
(b) The direct producers must be separated from their means of production in order that they can function as wage-labourers for the controllers of the means of production.

The questions immediately raised by this are:

(1) What are the origins of this accumulation of monetary capital—

how does monetary capital come to be accumulated in the hands of those who control the use of the means of production?

(2) Under what conditions do direct producers come to be separated from their means of production, in order that they can become wage-labourers for capitalist production?

(3) What theoretical concepts do we need to analyse these problems?

In answering these questions, we have to examine the genealogy of those elements that combine to form the specifically capitalist mode of production. As we will see, each of them has its own 'history', and this, of course, has to be uncovered if we are begin to examine the preconditions for the development and reproduction of a capitalist mode of production.

In order to facilitate this investigation, we can examine Marx's discourse, in *Capital* and the *Grundrisse*, where, on several occasions,[1] he looks at this problem of the preconditions for the existence of the capitalist mode of production, for the formation of capitalist relations of production.

Marx's analysis

In *Capital*, Volume I, Marx states:

> In themselves, money and commodities are no more capital than are the means of production and of subsistence. They want transforming into capital. But this transformation itself can only take place under certain circumstances that centre in this, viz. that two very different kinds of commodity-possessors must come face to face and into contact; on the one hand, the owners of money, means of production, means of subsistence, who are eager to increase the sum of values they possess by buying other people's labour-power; on the other hand, free labourers, the sellers of their own labour-power, and therefore the sellers of labour.[2]

These, then, are the basic conditions for capitalist production. From this basis, Marx then goes on, in both *Capital* and in the *Grundrisse*, to specify various historical 'routes' by which these preconditions can emerge.

The separation of direct producers from their means of production

In *Capital*, Marx takes England, and English feudalism as his example;

he shows how, during the period 1560–1700, the two processes of transformation of arable land into pasture land (for the embryonic woollen industry) and the selling of the church estates led to a massive expulsion of peasantry from their rented holdings; this was followed in the eighteenth century by enclosure, and then in the nineteenth century by the infamous 'clearing of estates' which intensified expropriations. For our purposes the precise course of these events is less important than the reasons that brought them about. What we want to know is how this historical process of separation of producers from their means of production was actually possible.

A major aspect here is that, from the very beginning, it was achieved largely through the intervention of state power within the feudal system of production. During the fifteenth and sixteenth centuries the political alliance on which state power was based increasingly became one between the new landed aristocracy (trading in the agricultural commodities they produced), the rapidly developing merchant groups (largely involved in the woollen trade), and the embryonic capitalist class (merchants who gradually gained control over the producers who supplied them). During this period, it was in the interests of the classes and fractions participating in this alliance that the peasantry should be expropriated – for obvious reasons. The expulsion of the peasantry from their lands and the subsequent production of commodities for the market enabled landowners rapidly to augment the surplus they received from production. This process coincided with the manufacturer's need for wage-labourers, which were abundantly provided by agricultural expropriation, perpetually increasing the numbers of unemployed, who could then sell their labour-power. Furthermore, the development of this labour force also created a home market for both industry and agriculture. In addition, expropriation of the peasantry enabled the embryonic forms of capitalist industry to compete more efficiently with rural-based artisan industry, since it gradually undermined the material basis for the latter's existence. We could continue with other examples, but this would only reinforce the point—that, in the example of English feudalism, the process of separation of direct producers was brought about through the coercive use of state power, through the intervention of the political level, an intervention based on the mutual interests of the different classes and fractions dominant within the social formation. The essential precondition for capitalist production, the separation of direct producers from their means of production, appears, from Marx's example, to result from a conjuction of two events: first, the transformation of agricultural production made possible by the expansion of the woollen trade in Europe and then in England, an expansion which was initially based on the development of the rural artisan woollen industry. This event then provided the impetus for the landowner's expropriation. Secondly, the absolute necessity for wage-labour on the part of the embryonic capitalist class. The important point is that both these events occurred under the dominance of a non-capitalist mode of production. Indeed, it seems that the articulation of the dynamics of the feudal mode of production within the social formation created the preconditions in which one of the basic elements of the capitalist mode of production could emerge. The origins of the separation of direct producers can, therefore, be traced within the transformation of the agrarian structures of the previously dominant mode of production.

From this example, it seems that the origins of the capitalist mode of production must initially be related to the dynamics of the previously dominant mode, *to the extent to which its articulation within the social formation* (evidenced in the use of the legal, military and political power necessarily involved in the reproduction of the feudal mode) *can establish preconditions for the emergence of the elements that capitalist production necessarily has to combine to reproduce itself*. In Marx's example, the development of the European woollen industry, together with the increase in the output of the English rural artisan industry, and the subsequent rise in the price of wool, led to conditions under which an increase in the amount of ground-rent extracted by landowners could (and did) take place. Thus, it seems that the surplus appropriated increased within the limits of the forms of ground-rent (labour-rent, rent-in-kind, or money-rent) in existence at this time under the dominance of feudal relations of production, and that this established the basis for a series of other effects—namely the augmentation of mercantile capital and the possibility for a political alliance between the classes and fractions mentioned above, at the political level. Thus the articulation of the dynamics of the feudal mode within the social formation established the basis for the existence of effects that could lead to a process of separation of direct producers. Yet, there is nothing inherently necessary about this process. The dynamics of the feudal mode do not, in themselves, as some writers have suggested,[3] require this. Whether or not these effects are realised depends upon the particular conjuctures in which they exist—upon such aspects as the nature of the political alliances established between the dominant classes and fractions; upon the outcome of the conflicts between this alliance and other landowning groups not producing raw material for the embryonic textile in industry, with sections of the peasantry, rural artisans, etc.; upon the political struggle between the feudal nobility and the monarchy over the utilisation of the locally produced surplus during the period of expansion of trade; upon the struggle within the nobility over legal rights to the surplus realised, and so on.[4]

If we now turn from this particular example to Marx's analysis in the *Grundrisse*, we see that, in contrast, other modes of production he examines *do not appear to allow the development of any separation of direct producers from their means of production.* The Asiatic mode, for instance, (as we shall analyse later in the text, and as Marx indicates both in the *Grundrisse* and *Capital*) has a particular form of dynamics which results in a necessarily limited development of the productive forces, in which a separation of direct producers on any large scale is impossible.

Before developing these points on the possibility for the separation of direct producers from within different non-capitalist modes any further, however, we should return to Marx's anaysis of the transition to capitalism, and briefly trace the genealogy of the other basic element of the capitalist mode of production, namely capital itself.

The accumulation of monetary capital

Marx raises this question at the end of *Capital*, Volume I:

> Whence come the capitalists originally? For the expropriation of the agricultural population creates, directly, none but great landed proprietors.[5]

The answer which he gives to this is that the accumulations of monetary wealth existing in the European transition from feudalism to capitalism had four origins: usury, mercantile capital, state deficit financing and (to a much lesser extent) hoarding by peasants and tenant farmers. The first three of these forms developed within the feudal formation alongside the exchange of commodities. This development, enabling the landowner to exchange his corn, cattle, sheep, etc., for other commodities on the market, depended upon and, in turn, extended, the role of the merchant. Similarly, mercantile capital was given a tremendous boost by the colonial trading of the fifteenth, sixteenth and seventeenth centuries. Again, usury (particularly usury on landed property but also in its many other forms[6]) rapidly became an essential institution. Similarly, the development of trading and the colonial system, with its concomitant commercial wars and inter-state trading relations, produced a complex system of international credits and financing, which gave ample opportunities for joint-state companies and banking institutions to insert themselves into the web of trading links and amass considerable sums of monetary wealth.

The actual historical development of these processes is, of course, well known, being rigorously documented in the works of economic historians such as Hobsbawm and Hill.[7] Two points from Marx's analysis need stressing, however, both because they are often overlooked, and because they are important for us here.

The first point is that *these forms of monetary wealth can arise in very different modes of production*; they are not confined, as many writers have suggested, to the transition from feudalism to capitalism. Monetary wealth can be accumulated as soon as products begin to circulate and exchange on the market through a medium of exchange, and this process can occur in various modes of production. For example, accumulations of monetary wealth existed in the Asiatic modes of S.E. Asia prior to colonial penetration. They developed in Latin American countries along with the development of commodity circulation.[8] Similarly, accumulations of barter goods pre-existed colonialism in Africa before they were turned into capital with the monetarisation of the economy during the colonial period. Consequently, it seems that whilst a separation of direct producers cannot emerge from the articulation of all non-capitalist modes of production, it nevertheless appears that accumulations of monetary wealth can exist within many different non-capitalist modes, once the circulation of commodities develops. This point, as we shall see shortly, is of crucial importance for our analysis.

Our second point is, effectively, a corollary of this. Namely, that accumulations of monetary wealth, which can arise through circulation in different modes of production, are not, in themselves, a sufficient precondition for the existence and reproduction of a capitalist mode of production. As Marx states in the *Grundrisse*:

> But the *mere presence of monetary wealth* . . . is in no way sufficient for this *dissolution into capital* to happen. Or else ancient Rome, Byzantium, etc., would have ended their history with free labour and capital, or rather begun a new history.[9]

The separation required for the reproduction of the capitalist mode

The capitalist mode of production requires a wage labour separated from ownership and control over the means of production it operates.

Although once established, the capitalist mode can reproduce this separation on an expanding scale, it *cannot initially be established without it.*[10] Consequently, unless the dynamics of a particular mode of production, in its articulation with the other instances of the social formation within which it exists, can establish the preconditions for a separation of direct producers from their means of production, capitalist production will not be able to emerge within the social formation. Any analysis of the history, or the origins of the capitalist mode of production must, therefore, necessarily trace the genealogy of the elements that constitute its structure. Furthermore, it is also clear that the historical field in which these elements are constituted is precisely that of the dynamics of the mode of production that pre-existed the dominant capitalist mode, of its articulation within the instances of the social formation in which it exists.

In our example, Marx investigates the genealogy of these elements from within a particular historical field—the English feudal formation—but he also indicates that other modes in other social formations may, as a result of their own processes of development, be unable to constitute these elements, even in an embryonic form. This provides us with an important conclusion.

We are attempting to analyse the preconditions for the existence of an articulation of two modes of production in a particular form of transitional formation. Within this articulation, the capitalist mode is becoming increasingly dominant. What we want to investigate, therefore, are precisely the preconditions for the existence of this mode of production. Hence, what we need to examine is the genealogy of the elements of the capitalist mode as they were formed within the social formation and mode of production that pre-existed capitalist production. We must initially trace the genealogy in the tendential development of these modes of production and the social formations in which they existed. This can be complex; the elements of the capitalist mode can, as we have indicated, be constituted in many different ways, under very different and independent historical conditions, and they always exist on the periphery, marginal to the social formation.

Having outlined the genealogy of the capitalist mode, we can now turn to the problem of analysing the modes of production that pre-existed the articulation of modes characteristic of Third World formations. How are we to investigate these modes of production, their structure, reproduction and tendential development, in relation to the genealogy of the capitalist mode? What theoretical concepts can we use in this investigation?

8 Theorising the Non-Capitalist Mode of Production: Problems and Perspectives

'IDENTIFYING' A MODE OF PRODUCTION

In the chapter on the object of historical materialism, we indicated that a mode of production exists as a double combination of invariant elements, and that this combination determines which of the practices will be determinant in the last instance in the social formation, in which this mode is articulated. Consequently, the solution to the problem we posed in the last chapter would seem to be relatively simple.

Examining the economic systems of pre-capitalist societies, we simply have to specify the forms in which these elements combine and the nature of their double combination. Alternatively, we can apply what we consider to be the combination appropriate to the Asiatic or Feudal modes to particular societies. Such an approach is revealed, for example, in the work of Emmanuel Terray,[1] who, analysing the lineage mode of production, argues that to construct a theory of the mode or modes of production operating in a given social formation, one has to identify or make 'an inventory',[2] of the labour processes that exist; one then classifies them in terms of how they combine the elements, labourer, means of labour, object of labour, and non-labourer, and on this basis, define the dominant relations of production, distribution, and the representation of these relations in political and ideological forms. Such a procedure begins by locating processes of production; modes of production are then defined by discovering pre-existing theoretical elements in these processes. Terray's analysis of Gouro society in *Marxism and 'Primitive' Societies*,[3] for example, is founded on a specification of the instruments of labour utilised, which act as indices, a starting point for the construction of the particular economic base specific to the lineage mode. Modes of production are thus constituted by locating a combination of elements in a description of a process of production.

Much the same conclusions can be reached regarding the analyses of modes of production carried out by Meillassoux,[4] or the analyses of the Asiatic mode of production carried out by the authors of the C. E. R. M. Group,[5] who apply 'fully-formed' concepts of modes of production from Marx's discourse.

In each case, the procedure is the same—identify the combination of elements that appear to exist in the process of production, and base one's analysis of the mode of production on it. Such an approach contains a number of errors which we must briefly examine, since their pervasiveness in the analysis of non-capitalist societies is a barrier to be overcome, if the concept of mode of production is to be used in the theorisation of Third World formations.

As we pointed out earlier in the text, any mode of production exists as a *structure in dominance*, as a combination of different labour processes under the dominance of a specific mode of extraction of surplus labour, determined by the dominance of particular relations of production. In other words, in the double combination of elements in the mode of production, it is the relations under which the surplus labour is extracted that are dominant; it is they that structure the processes of production. What this means is that the analysis of any mode of production cannot simply be approached by identifying elements and their combination in the process of production, the division of labour, since the process of extraction of surplus labour can never be *derived from* the process of production, precisely because the latter depends on, is structured by, the former. Thus the dominant relations of production, the relations existing between the labourer and non-labourer, which are the axiomatic basis for analysing the specificity of any particular mode, will remain hidden if we confine our analysis to the level of the production process. This is an error that very many theorists of non-capitalist social formations have fallen into; an error which has led them either to generalise the application of a particular mode of production to such an extent that its specificity becomes totally lost,[6] or to conflate relations of production into relations of appropriation, thereby misrecognising the dominant relations of production and the social division of labour that is the very basis for the class structure itself.

This, then, is the first 'pitfall' to be avoided in analysing pre-capitalist modes—to avoid the fallacy of simply identifying a combination of preexisting theoretical elements in the process of production.

Having said this, however, our task then becomes much more difficult, since what we have to uncover are those relations of production which permit surplus labour to be extracted from the direct producers in the particular mode we are analysing. Marx, of course, did this when he analysed the relations of production that governed the extraction of surplus labour in the capitalist mode of production, or, again, when he specified the relations under which ground-rent was levied from the direct producers in the feudal mode of production. In both cases, 'the specific economic form in which surplus labour is pumped out of direct producers'[7] was never given in the immediate process of production. Rather, the relations of production that determined this form had themselves to be constructed, not by examining the process of production, but by beginning to analyse the overall reproduction of the total economic system in relation to the social formation itself. This, then, is the task we face in beginning to analyse non-capitalist modes of production. We must begin by examining production in relation to its other moments—circulation, distribution, etc.—and in its articulation with the other levels. Only then can we pose the problem of the existence of the particular relations of production that determine this system of production, which enable it to reproduce itself, and which provide the basis for the social division of labour on which its class structure is based. In all cases, we must initially approach the economic structure as a system, examining its articulation within the social formation, and bring the basic concepts of historical materialism to this rather than simply applying them to different processes of production, where we think we can 'read off' the particular combinations (Asiatic, feudal, etc.) that we have already constructed in theory.

In addition to 'identifying' a mode of production, there is a further error which many authors commit when they begin to approach non-capitalist formations.

THE LIMITATIONS OF THEORETICAL EXTRACTION

The fallacy consists in extracting the different concepts that Marx used in analysing non-capitalist formations at different periods of his work without adequately situating them in his discourse. Such an approach necessarily avoids the crucial problem of the *theoretical adequacy of these concepts* in analysing particular non-capitalist formations. Are the concepts set to work in the *German Ideology*, more appropriate, for example, than those operative in the *Grundrisse*, or, indeed in *Capital* itself? These questions have to be posed and answered before we go any further, since most analyses either assume that the concepts in these different texts are all appropriate for their object, or they simply operate with concepts developed in one text without assessing the adequacy of these in relation to their use and development in another text. Here we can only present a brief outline of a problem that will have to be taken up more rigorously if the confusions that many theorists in this field have fallen into up until now are to be avoided.

The first texts in which Marx deals specifically with non-capitalist formations are the *German Ideology* and the *Manifesto*. Both these texts are written at a point when, as Marx himself says, 'we resolved . . . to settle accounts with our erstwhile philosophical conscience', at a point where, having rejected the philosophical synthesis of Hegel and Feuerbach contained in the 1844 Manuscripts, Marx begins to outline what he terms a 'science of history'. These texts announce the project to be pursued in the *Contribution to the Critique of Political Economy*, and in the *Grundrisse*, which is culminated in *Capital* itself. Consequently, they do not establish an adequate basis on which we can analyse non-capitalist modes, as is clear from the texts themselves.

The *German Ideology* is an inherently evolutionist work. It reduces different social formations to effects of developments in the division of labour, which progresses as a result simply of the growth of the productive forces. This movement is said to produce a linear transition through different forms of property—tribal, state, feudal—to capital itself. The tremendous distance of this text from the theory of the structure and reproduction of modes of production formulated in *Capital* should be obvious. As opposed to the specification of their forms, their tendential development, their determinancy of the structure of the social formation, and the possibility of theorising the transition from one mode to another, we are presented with an evolutionist typology in which none of these problems—which are crucial for us here—can be posed. The *German Ideology*, then, can never be taken as the basis of any analysis of non-capitalist modes, since it indicates the site, the necessary future direction of Marx's work, rather than constructing a corpus of theoretical concepts appropriate to the analysis of the structure and reproduction of modes of production.

Much the same can be said for Marx's series of articles on aspects of British colonialism for the New York *Daily Tribune* from 1953–7.[8] Although they provide us with valuable insights into how commodity-export undermines the reproduction of non-capitalist modes of production, and how colonial rule can transform agrarian structures in line with the reproductive requirements of the capitalist mode, they do not advance theoretically beyond the formulations contained in the *German Ideology*.

The *Grundrisse*, and particularly its section, 'forms that pre-existed capitalist production', is taken by many theorists as constituting Marx's most exhaustive statement on pre-capitalist economic formations. Yet this conclusion is highly questionable, for two reasons.

First (and less importantly) because the object of the text is not specifically an analysis of the structure and reproduction of non-capitalist modes, or the transition from one mode to another, but rather an analysis both of the preconditions for the existence of the capitalist mode of production—thought on the basis of the structure of this mode—and a survey of the different historical conditions by which these preconditions can be constituted. The object is to account for the relatively independent histories of the two axiomatic elements which combine to establish the basis for capitalist production and reproduction. Thus, the different accounts of the various pre-capitalist forms only enter into the text to the extent that they are essential for this overall objective. This consequently limits any analysis that tries to base itself on this text. However, before going on to indicate the effects of operating with the concepts developed in the 'Formen', we should briefly turn to the second (and more fundamental) limitation that is placed on using the *Grundrisse* as the foundation text for the analysis of non-capitalist formations.

The text constitutes a draft essay on political economy, a draft of Marx's project for a text on capital encompassing an outline which was then considerably expanded and transformed in *Capital*[9] itself. The *Grundrisse* is a draft written at a particular moment in the development of a theoretical discourse which culminates in *Capital*. Hence it is essentially a transitional work, a text written at a point where the concepts fully developed in *Capital* were in a state of gestation. This is clearly evident in the text. For example, concepts developed during Marx's Feuerbachian period are used to think problems that are posed in totally different conceptual terms in *Capital*. The transition from one mode of production to another is understood in the *Grundrisse* through the Hegelian concept of negation, the concept of alienation is used to analyse the role of money in the exchange of commodities, and so on. Concepts that are of crucial theoretical importance for historical materialism, such as the 'contradiction between the productive forces and relations of production', or the determinancy by the mode of production, are only indicated as problems, and remain undeveloped.

Other key areas of historical materialist analysis of the mode of production are simply absent; for example, there is no rigorous theory of reproduction (which tends to be explained via a notional use of 'alienation'), nor is there a theory of crises (which are generally explained as being a result simply of 'overproduction'); again, the overall structure of the mode of production itself remains untheorised. In addition, there is no fully elaborated theory of surplus-value, and the problem is often posed in Ricardian terms, which Marx later rejects. Consequently, the *Grundrisse* is a text in which the concepts of historical materialism, fully presented in *Capital*, are still being developed. As such, it bears all the hallmarks of its transitional status—the sites of problems are indicated, but, in many areas, they remain unresolved. Concepts are used to specify problem areas, but are later cast aside as being inadequate for the phenomenon they presumed to explain.

These two problems of the specific object of the 'Formen' section and the theoretical status of the *Grundrisse* place severe limitations on the use of this text in analysing non-capitalist social formations. To make this more specific, let us take two examples.

First, the problem of transition from one mode of production to another. In *Capital*, whilst we have no coherent, fully worked out theory of transition, we are given a specification of the dynamics of the capitalist mode of production, and a theory of the genealogy of the elements of this mode. This provides us with a foundation for analysing transitional social formations, which is clearly not the case with the 'Formen'. Here, as opposed to any specification of dynamics, we are given a formulation in which the extended reproduction of a mode of production, in itself, is said ultimately to engender the dissolution of this mode; that is, the internal reproduction of the structure necessarily generates 'transition'. Thus, transition is simply conceived in evolutionist terms, as somehow being a result of production; for example:

> The survival of the commune as such in the old mode requires the reproduction of its members in the presupposed objective conditions. Production itself, the advance of population (this too belongs with production), necessarily suspends these conditions little by little: destroys them instead of reproducing them etc., and with that, the communal system declines and falls, together with the property relations on which it is based.[10]

Consequently, if we want to analyse the *process* by which, during a transitional period, the structure of the mode of production begins to be

formed under the dominance of another mode, or how one mode can penetrate and articulate itself within another, this formulation is of little value, since, as opposed to any analysis of the dynamics of different modes, of their tendential development, all that is stated is that transition is somehow the result of 'overproduction', of one form 'overproducing' itself, and giving way to another. The transition, then, according to this formulation, simply happens, and we have no adequate basis for analysing why or how.

Similarly, the *Grundrisse* provides us with no concept of mode of production. As opposed to *Capital*, which specifies the invariant elements of modes of production and their different forms of combination in various modes, the *Grundrisse* simply describes different forms, largely on an empirical basis. Consequently, any attempt to found a theory of non-capitalist modes from within the *Grundrisse* usually ends up in some sort of stages theory, with broad descriptions of different socio-economic forms which generate one another in rather haphazard ways.

Again, because of the absence of any theorisation of the determinancy in the last instance by the mode of production, of the way in which the latter structures the other instances in the reproduction of the social formation, the *Grundrisse* is equally limited when we pose to it the problem of the concept of social formation. Indeed, as opposed to *Capital*, where these structural articulations are clearly specified, the *Grundrisse* tends to analyse ideology, in particular, as being simply a form of estrangement—again via the use of 'alienation'.

We could continue with further examples, but this would only be to labour the point. Unless the object and theoretical status of the 'Formen' section of the *Grundrisse* is clearly posed, we will continue to operate with concepts that are inadequate to their object. Whether we are examining the structure, reproduction, dynamics, genealogy or articulation of modes of production, the concepts outlined in the *Grundrisse* are severely limited.

When we turn to *Capital*, with its theoretical specification of these problems, we can clearly see the limitations of the theoretical discourse of the *Grundrisse*. Only in *Capital* do we find the concept of mode of production—the theory of the invariant elements of all modes, an exposition of their differential combinations in particular modes of production, and an account of the concept of determinancy in the last instance by the mode of production. Furthermore, only in *Capital* are we presented with a coherent theory of reproduction—with an analysis of the general formal conditions for the reproduction of modes of production ('simple reproduction') and a particular theory of enlarged reproduction and accumulation within a capitalist mode of production. Again, only in *Capital* are the concepts of dynamics and genealogy developed and coherently thought out on the basis of an analysis of the structure and reproduction of modes of production. With all these concepts, Marx is, of course, primarily concerned with the capitalist mode of production, but, nevertheless (as we hope to show below), they can be utilised in analysing the structure, reproduction and dynamics of non-capitalist modes of production.

To exemplify how these concepts can be used in building a more adequate framework for analysing the modes of production that pre-existed capitalist penetration, I intend briefly to examine the structure and development of one particular capitalist mode that pre-existed capitalist penetration in S. E. Asia—the well known (and much misunderstood) Asiatic mode of production. Before going on to this, however, it is essential to criticise two further approaches to the use of the concepts of historical materialism.

The first is formed by a reading of these concepts, which, following contemporary usage, we can term 'structuralist'. This reading reduces the transition from one mode of production to another to a variant of evolutionism, in which the transition is produced solely by the development of the dominant mode of production. Such an approach—which is increasingly popular in the analysis of non-capitalist modes of production[11]—effectively leads us back into the errors in analysing the articulation of modes in a transitional period that we have located in the *Grundrisse*. To demonstrate this, I intend to examine the work of Maurice Godelier, perhaps the best known structuralist reader and appropriator of Marx's writings on non-capitalist social formations.

The second approach is best described by the term 'formalist', since it is based on a reading of Marx's texts which requires a formal correspondence between sets of relations and forces of production as a pre-condition for theorising a mode of production. This reading is contained in the work, *Pre-Capitalist Modes of Production*, by Hindess and Hirst.

Taking these approaches in turn.

THE STRUCTURALIST ANALYSIS OF NON-CAPITALIST MODES: THE WORK OF GODELIER[12]

Godelier's concept of mode of production and its limitations for theorising transition, are perhaps best exemplified by returning to one of his early articles, 'System, Structure and Contradiction in Capital',[13] where he outlines the structure of the capitalist mode of production, arguing that it contains two contradictions—one 'internal' to the relations of production between capital and labour, and the other 'external' to these relations, between the forces and relations of production. Whilst the former contradiction is original to the structure of the capitalist mode, the latter contradiction only develops as the basic contradiction at a particular stage in this mode's history:

> Thus the basic contradiction of the capitalist mode of production is *born* during the development of the mode of production, and is *not present* from the beginning of the system.[14]

Furthermore, Godelier argues, the appearance of this contradiction indicates a limit to what he terms the 'functional compatibility' of the two structures, the relations and forces of production.

> It [the basic contradiction] signifies the *limits* within which it is possible that capitalist relations of production, based on private property, may correspond to the development of the productive forces to which they have given birth . . . . they are thus limits expressing objective properties of the capitalist mode of production.[15]

With the development of the capitalist mode, the relations of production under which the productive forces have been developing, become 'dysfunctional', and 'irrational' for the further development of these forces. In Godelier's structuralist terminology, the diachronic development of the synchronic capitalist structure creates a situation in which the relations of production block the development of the productive forces. It is on this basis that Godelier argues that:

> Thus, while ceaselessly developing the productive forces, capital *unconsciously* creates the *material* requirements of a higher mode of production, and *necessitates* [our stress] the transformation of capitalist conditions of large-scale production based on private property into 'general common social conditions'. The development of capitalism makes possible and necessary the appearance of a socialist economic system, of a 'higher' mode of production.[16]

Thus, for Godelier, a 'socialist mode of production' is generated as a primary effect of the linear temporality of the capitalist mode of production. Capitalist production must engender a socialist economic system, because the development of the productive forces requires it; and, as opposed to the 'dysfunctioning' of the productive forces under the dominance of capitalist relations,

> . . . the *structure* of socialist relations of production *corresponds functionally* [our stress] with the conditions of rapid development of the new, gigantic, more and more socialised forces created by capitalism.[17]

From this general outline, it would appear that, for Godelier, the 'basic contradiction' appears at a certain stage in the diachrony of the capitalist mode (when capitalist relations begin to retard the productive forces), and the auto-genesis of this contradiction, *in itself, produces a transition* to another mode of production. The diachrony of the structure sets limits to its own development, and once these limits are reached, a transition to another mode of production occurs as a necessity.

This overall approach in analysing the structural development of the capitalist mode became the one that Godelier adopted for analysing the possibilities for transition in non-capitalist modes of production. To Lévi Strauss's conception of the diachrony as a sequence of historical events generated by the synchronic structure, Godelier added an element of historical necessity, and reduced the dynamics of the mode of production and the transition from one mode to another to an inexorable deterministic evolution of one fundamental antagonism, between the 'relations and forces of production'. The analysis of history thus came to be approached via a mechanistic interpretation, in which all events are explained by reducing them, as G. Sofri aptly states, to 'the solution of the contradictions inherent in chronologically antecedent forms'.[18] From such a position, Marx's analysis becomes a form of economic fatalism, in which the analysis of the logic of a structure enables us to define its possibilities and its potentialities for evolution.[19]

The effects of Godelier's concepts for the analysis of transitional formations

What, then, are the effects of Godelier's conception of transition for analysing the particular transitional formations we are concerned with here? As we indicated above, a mode of production must be analysed as a structure in which the relations of production dominate the productive forces, in which the former impose a particular development on the latter. Consequently, rather than basing our analysis on the 1859 Preface (where Marx outlines the historical field he is about to develop), and asserting that there is a 'dysfunctional antagonism' between the relations and forces of production, it is essential that we begin to analyse *the particular patterns of development that specific relations of production impose on the productive forces in different modes of production*. This point is vital for analysing transitional formations, since we can only begin to analyse these periods if we can specify the dynamics of the previously dominant mode of production and how the articulation of these dynamics in the social formation can (or cannot) establish possibilities for the emergence of the elements of a new mode of production.

Yet, Godelier ignores all this. For him, it is simply a question of the development of a *presumed contradiction necessitating transition*.

Take, for example, his introduction to *Sur les Sociétés pré-capitalistes*,[20] and in *Sur le Mode de production Asiatique*,[21] where he indicates that the Asiatic mode of production develops as a result of the 'evolution' of a contradiction inherent to kinship systems in hunting and gathering or pastoral societies, and how the Asiatic mode can evolve into slave or feudal societies, as a result of a 'contradiction' inherent to its own structure. For example, in *Sur le Mode de production Asiatique*, Godelier states:

> This power [the power of the state over the communities in the Asiatic mode] originally takes root in the functions of common interest (religious, economic, political), and is gradually transformed into a power of exploitation without ceasing to be a power of function. . . . The very essence of the Asiatic mode is the combined existence of primitive communities, in which common and organised possession of the soil exists, still partially on the basis of kinship relations, and a *state power* which expresses the real or imaginary *unity* of these communities, *controls* the use of essential economic resources, and *directly appropriates* a part of the labour that it dominates.

The reproduction of this structure, in itself, produces transition:

> For us, the law of evolution of the Asiatic mode—as for every other social formation—*is the law of development of its internal con-*

> *tradiction.* The internal contradiction of the Asiatic mode of production is that of the unity of community structures and class structures. The Asiatic mode of production evolves through the development of its contradiction towards forms of class societies in which the community relations are increasingly inappropriate due to the development of private property.[22]

Thus, rather than specifying the structure of a mode of production, or (on the basis of this) analysing the subsequent patterns produced by the historical process of subsumption of the productive forces under particular relations of production, and the effects this has on the other instances, Godelier reduces this dynamics to the auto-development of the structure. Consequently, his notion of the contradiction between relations and forces of production is one in which the process of subsumption of the productive forces under the dominant relations of production, and the economic effects this sets up, cannot be thought. Any analysis of the dynamics of the previously dominant mode of production is ruled out of court at the outset.

In addition to this, Godelier's notion of a mode of production also severely limits the analysis of the relationship between the economic and the other instances in transitional formations, of the way in which the articulation of the dynamics of the mode within the social formation can establish the possibilities for transition.

This is exemplified in his recent analysis of the Mbuti tribe in 'Anthropologie et économie', where he puts forward the following formulation:

> When we analyse closely these economic and social relations, we see that the very conditions of production determine three constraints within the mode of production, and that these constraints explain (traduisent) the conditions for the *reproduction* of this mode of production. These three constraints express the social conditions for the reproduction of the process of production, given the nature of the productive forces put into operation.[23]

Here, instead of any structural analysis of the mode of production and its concrete articulation with the other instances, we are given a rather vague formulation in which the level of development of the productive forces itself determines both the relations of production and social relations, imposing functionalist constraints on them. Godelier argues, for example, that since the hunting and gathering tribe must be fairly widely dispersed over an area, and their hunting must move from one location to another, this factor, together with the sexual division of labour required for hunting and gathering, necessitates particular kinship relations both within the tribe and with other tribes—kinship lines and kinship alliances must all be in a state of permanent 'flux' within and between tribes; there must also be a strong emphasis on sexual and generational differences, and a regular means of exchanging women between tribes. Thus, the constraints imposed by the conditions of production, themselves determined by the development of the productive forces, express the conditions for the reproduction of the social formation. Phenomena such as kinship, religion, etc., are analysed as *'functional necessities' in relation to the level of development of the productive forces*:

> Thus, these constraints, *internal to* the mode of production, are at the same time the channels by which the mode of production determines in the last analysis the nature of the various instances of Mbuti society . . .[24]

The basis of Godelier's concept, then, is clear. Rather than the structure of the particular combination of elements in a mode of production ultimately determining and structuring the other instances, the latter can only be analysed as effects of the functional needs of the productive forces; the labour processes required for material appropriation constrain the development of the other instances, regardless of the relations of production and the social division of labour which in all modes of production dominate and structure these processes. Consequently, rather than taking the structure and reproduction of the mode of production as the foundation for his analysis, Godelier concentrates exclusively on the productive forces, on the labour processes in isolation from the relations of production which govern them. He never analyses the latter, and, as a result, he simply derives the relations of production from the nature of the labour processes, rather than from an analysis of the process of extraction of surplus labour that enables the mode of production itself to be reproduced. The fact that the notion of 'constraints' becomes a substitute for analysing the process of articulation of instances as structured by the determinant instance is a direct result of this error.

The transition from the dominance of one mode of production to another exists as a conjunction of different processes: the dynamics of the previously dominant mode of production, the constitution of the elements of the newly emerging mode, and the possible articulation of other modes of production in the social formation to ensure the establishment and/or reproduction of the new mode. It is only by analysing the inter-relation of these processes in the transition period that we can establish a framework for examining the historical development of Third World formations, since it is these very processes that ultimately determine the structure and tendential development of these formations. We have tried to briefly indicate how Godelier's structuralist reading of the concept mode of production cannot provide a basis for adequately theorising these processes. The evolutionist perspective that results from his reading prevents them being conceptualised in any rigorous form.

THE FORMALIST READING OF NON-CAPITALIST MODES OF PRODUCTION

Subsequent to the publication of *Pre-Capitalist Modes of Production*[25] (*PCMP*), its authors have criticised the theoretical framework it put forward, arguing that it was based on a rationalism which they now regard as untenable. In *Mode of Production and Social Formation*[26] (*MPSF*), they conclude that their earlier work required a relation between its theoretical discourse and reality in which the latter was reproduced through the order of the discourse. The concepts of modes of production, developed in theoretical discourse, determined and structured reality. The connections between real objects were determined by the logical ordering of conceptual connections in theory. In short, phenomena were analysed as forms whose relations were simply equivalent to their concept. In *MPSF*, it is argued that a knowledge of objects within a theoretical (or any other) discourse cannot provide a knowledge of objects as they exist in reality; since one discourse can only be assessed in relation to another discourse, there cannot be any epistemological legislation on the adequacy or otherwise of discourse in relation to reality. Relations between real objects and objects of discourse cannot be conceptualised, since real objects cannot be represented in discourse (and vice versa); any attempt to do so, whether it be empiricist, positivist, idealist, or—in the case of *PCMP*—rationalist, is necessarily based on epistemological protocols whose validity can only be established by reference to a further discourse. Thus, *MPSF* is not denying that objects exist beyond discourse, but it is denying that their existence can take the 'form of objects representable in discourse'.[27]

Elsewhere, I have argued that Hirst and Hindess's characterisation of Althusser's epistemology as a rationalist philosophy of history is untenable.[28] From this, it should be clear that I also consider it possible to have a theory of modes of production and social formations that is not limited by the notion of rationalism in which Hirst and Hindess have confined it. In addition, although this present text is not concerned directly with these problems, it should be apparent from it that theoretical discourses on ideologically received objects can be assessed in relation to crucial aspects of these objects which they include, exclude, and can or cannot theoretically pose problems about; that these discourses can also be assessed in terms of their closeness (or otherwise) to ideologically received notions as a whole—as to what extent they can go beyond these surface appearances to conceptualise the structure, reproduction, and future possible development of particular social formations. Consequently, the reasons put forward by Hirst and Hindess for rejecting *PCMP* are, it seems to me, questionable. Yet, whilst I disagree with these reasons, I concur with the rejection—but on the basis of a very different critique. Since the notions put forward in *PCMP* are being increasingly used in theorising non-capitalist modes of production, it is essential to outline here the main aspects of this critique. It concerns the particular concept of mode of production in *PCMP*, the effects of its notion of the articulation of the mode within the social formation, and its comments on the theory of transition from one mode to another.

Mode of production as a unitary articulation

The object of *PCMP* is clearly specified in its introduction:

> For each of the pre-capitalist modes of production briefly indicated in the works of Marx and Engels, we may pose the following question: 'Is it possible to construct the rigorous concept of that mode of production as a determinant articulated combination of relations and forces of production?' We proceed in each case to answer the question either by constructing the necessary concept or by demonstrating that no such construction is possible. If the concept can be constructed in accordance with the general concept of mode of production, then it is a valid concept of the Marxist theory of modes of production. Otherwise the concept has no validity in terms of Marxist theory.[29]

In each case, having posed the question of what constitutes a mode of production, and answering, 'An articulated combination of a specific mode of appropriation of the product and a specific mode of appropriation of nature',[30] the structure and reproduction of each mode is examined to see if an articulation of relations and forces *specific to it* can be formed from within the general theory of modes of production; if not, then the existing notion of the mode can be rejected. In other words, if a particular form of labour process can be shown to depend specifically on a definite set of relations of production, and if both process and relations determine each other in a structure whose determinancy ultimately lies with the relations of production, then a mode of production is formable. If there is no such unitary articulation, then it cannot be formed.

This conception has a number of limitations. In any one mode of production, as we have seen,[31] the relations of production define a set of positions for agents of production in relation to the principal means of production. These 'positions' are places where particular functions—appropriation, co-ordination, etc.—are carried out. In each mode, we have relations of production structuring a number of labour processes that embody these functions. The development of these processes is governed by the requirements for a specific mode of extraction of surplus labour. In the feudal mode, for example, there can exist forms of labour process variously based on co-operation, labour-service, tenant cultivation or even wage-labour. Similarly, in the lineage mode,[32] surplus labour is extracted by the elders from a combination of labour processes—the collectively organised system of hunting, production based on sexual and social divisions of labour in the extended family unit, the individual labour of slaves, etc. This structuring of different labour processes occurs when any mode of production becomes dominant in a particular social formation, and it is this, rather than an articulation of one set of productive forces with relations of production, which must be capable of being incorporated within the concept of mode of production. Yet, the text of *PCMP* can only establish the existence of a mode of production when there is a unitary correlation of two sets of relations and forces of production. This limits the analysis, since particular relations of production can occupy the place of determinancy in the last instance, whilst structuring different labour processes; we can also have a dominance of certain processes over others, in terms of the degree of their transformation by the relations of production. If we begin, however, by requiring an articulation of a labour process with a form of extraction of surplus labour, then this problem of the interrelation of labour processes under the form of extraction must remain untheorised, and an important aspect of the theory of the structure and reproduction of the mode of production must necessarily be omitted.

This is an important point, since the existence of any labour process is not simply deducible from the dominant relations of production: the latter set limits on the variation of the labour processes—which may well be those required by other relations of production—at different moments in the mode's development, as they become increasingly subsumed under the dominant relations of production. This is a more complex process than can be encompassed by the concept of a unitary articulation.

To illustrate the limitations of conceptualising modes as unitary articulations, we can select one particular example from the text of *PCMP*—that of the 'primitive communist mode of production'.[33]

Beginning with a definition of this mode as a collective appropriation of surplus labour in which there are no classes, state or politics, the authors develop a more specific formulation, characterising it as 'appropriation through the mechanism of the redistribution of the product.'[34] This mechanism, which governs the productive forces, has two variants—'simple' and 'complex' redistribution. In each case the relations of production are formed through different types of kinship. In the 'simple' variant, the product is redistributed through a network of relations established on a 'temporary or semi-permanent basis'. Conversely, in 'complex redistribution', the product is distributed through:

> . . . a stable and permanent network of relationships that is established in advance of any particular labour process or any temporary formation of groups.[35]

For each variant, there is a corresponding set of productive forces:

> The domination of simple redistribution therefore involves an extremely limited development of the productive forces, a rudimentary division of labour between sexes, and a mode of subsistence based on hunting and gathering.[36]

In contrast, under complex redistribution, a more systematic division of labour and extensive co-operation on a permanent basis are possible. Complex redistribution enables the formation of production communities with more 'refined' co-operative labour in herding and horticulture. Because it brings together more labourers and enables an increase in productivity, complex redistribution increases surplus labour time, which in turn permits the community to support specialised artisan production, an increase in the non-productive population, and the development of 'ceremonial and other activities, not directly connected with immediate production'. These phenomena, resulting from the development of the productive forces, provide the basis for 'a wider network of relationships'.

Turning to the reproduction of these two variants, the analysis is somewhat ambiguous. For 'simple redistribution' to be reproduced, for the surplus to be extracted and distributed, kinship relations are essential; yet, the authors also argue that, since there is no necessity for the perpetuation of social relations within and between bands, kinship is only a 'possibility'. For 'complex redistribution', the authors again state that 'the reproduction of the economy requires the reproduction of a determinate system of kinship relations',[37] since the co-ordination of production requires hierarchical relations within and between lineages and villages. Yet the analysis here remains descriptive, since the determinants of the reproduction of this economic-ideological articulation are never specified—rather, its importance for economic reproduction is simply asserted.

Having reached this point, we can now draw out the main steps in the author's argument.

Noting the major characteristics of the primitive communist mode, the authors specify two sets of social relations: from these they deduce two corresponding forms of redistribution, indicating that they require two different types of division of labour. They then argue that these necessitate sets of relations. For example: 'The dominance of complex redistribution provides the economic conditions of existence of a wider network of relationships.'[38] The tautology of this proof is clear: specific sets of social relations produce modes of real appropriation, which themselves then produce these sets; furthermore, the tautology rests on a description of different forms of organisation, rather than on a theorisation of the structure and reproduction of these forms and their determinants. The requirement for a unitary articulation of forces and relations of production has thus prevented a number of crucial questions being posed—concerning reproduction and its determinants, the possible limits of variation of the labour processes under the relations of production, the nature of the inter-relation between the economic and ideological instances, and the importance of this inter-relation for the reproduction of the mode of production. The concept has structured the analysis in such a way that it is necessarily restricted.

Structure and transition

As is well known, *PCMP* argues that, since a mode of production requires an articulation of the instances of the social formation which will guarantee its reproduction, it must be conceived as a self-regulating mechanism which perpetually produces its conditions of existence as an effect of its structure; as a result it becomes impossible to theorise transition from within this concept.

> . . . the conceptualisation of the mode of production in the mode of structural causality precludes any conception of transition from one mode of production to another.[39]
> If, say, the feudal mode of production reproduces its own conditions of existence, then how is transition to be conceived; if transition is to be possible, then how can each mode of production be considered as an eternity?[40]

If this argument were correct, it would, of course, become impossible to theorise the emergence of phenomena making possible a transition to dominance by a mode other than that dominant in the social formation at the moment of analysis. Analysing the possibilities for transformations in the economic and political structures of Third World formations would be severely restricted; but how valid is the authors' conclusion?

Certainly, the mode of production must produce—through its process of reproduction—an articulation which guarantees its existence, but why cannot this process have political and ideological effects that establish a possible basis for a transition to another mode of production? We have already indicated how the articulation of a feudal mode of production can produce a political effect which can begin to undermine the non-economic mechanisms ensuring its reproduction.[41] Why, therefore, cannot the articulation of a mode of production in a social formation produce transitional tendencies whose perpetuation depends on the development of conjunctures emerging as a result of the effects of these tendencies on the various instances? It seems to me that the analysis in *PCMP* requires the concept of mode of production to generate an explanation which it is not its theoretical function to produce. The authors state, 'Nothing in the concept [of mode of production] can entail its dissolution.'[42] But, one can ask, why should it since the preconditions for the emergence of transitional tendencies are never simply the result of the dynamics of the mode of production itself?

As we have indicated it is, rather, a question of the articulation of these dynamics within the times of the other levels of the social formation, developing unevenly, which creates these preconditions. As Balibar states:

> Indeed, if the effects within the structure of production do not by themselves constitute any challenge to the limits [of the existing mode of production], they may be *one of the conditions* . . . of a *different result*, outside the structure of production; it is this other result that Marx suggests marginally in his exposition when he shows that the movement of production produces, by the concentration of production and the growth of the proletariat, one of the conditions of the particular form which the class struggle takes in capitalist society . . .[43]

As with their concept of mode of production, the notion of mode of production as an eternity perpetually producing its conditions of existence, restricts the use of the theory of modes of production in an area which is crucial for the analysis of Third World formations. A theory which cannot adequately analyse the emergence of possible transitional tendencies is lacking in the very area—of assessing and predicting structural change—which should be one of its strengths.

Transition as a form of non-correspondence

Throughout the analysis of *PCMP*, the notions of non-correspondence or correspondence of the relations of production are equated with transition and non-transition. Thus, the critique of Balibar, for example, argues that he defines social formations in transition from dominance by one mode of production to another as characterised by a non-correspondence, and social formations dominated by one mode of production as characterised by a correspondence of relations and forces. This critique also enables the authors to attribute a teleological causality to the theory of transition in general, arguing that transition must necessarily be conceived teleologically, by the necessity to achieve an end-state (a specific mode of production). Thus, in addition to the structurally eternalist theory for non-transition we examined above, we now have a teleological theory for analysing transition. Yet, in order for this reading of transition as a teleology to exist, the initial premise of correspondence/non-correspondence = non-transition/transition has to be accepted. This, however, is very much open to question.

As we noted earlier, transitional social formations are characterised by an articulation of relations of production specific to (at least) two modes of production, in which the mechanisms (ideological, economic, political) ensuring the reproduction of one of these sets of relations are becoming dominant over the mechanisms ensuring the reproduction of other sets. The labour processes existing under the previously dominant relations of production begin to be transformed in directions required by subsumption under the newly emerging form of extraction of surplus labour. In the transition period there is, therefore, a non-correspondence between relations and forces of production appropriate to different modes. Yet, when the newly emerging relations of production become dominant, there is no reason why this non-correspondence cannot continue to reproduce itself, since the pre-existing labour processes cannot be easily transformed by the structuring action of their newly determining relations of production. As we shall see in a subsequent chapter,[44] despite the dominance of the capitalist mode in Third World formations, its relations have been unable to transform existing labour processes in several sectors. This is important for the analysis of these formations, since it affects both the form of capitalist development that emerges, and has important results for the class structure.

It would seem, therefore, that to equate transition/non-transition with correspondence/non-correspondence is a considerable over-simplification. Transitional periods are characterised by a fairly complex articulation of relations of production at the economic level, which themselves require the creation and transformation of very different labour processes. Similarly, when a particular mode of production becomes dominant in a social formation and there is no longer an articulation of different relations, its relations of production can still structure labour processes to which they do not simply 'correspond'. In either case, it would seem that the analysis of the relations between labour processes and relations of production cannot be confined with the correspondence/non-correspondence dichotomy.

In our assessment of the framework for analysing modes of production put forward in *PCMP*, we have focused on particular aspects of the text—on the adequacy of its concept of mode of production, on its strictures for analysing the possibility for transition with a social formation, and on its critique of the analysis of transitional periods based on the theory of modes of production.

Our critique has focused on these because they are crucial problem areas for the theory of modes of production in its analysis of Third World formations. In each area, it seems that the notions developed in *PCMP* would restrict the analytical field of the concepts to such an extent that their level of generality would render them untenable.

9 Conceptualising Non-Capitalist Modes of Production: the Asiatic Mode

In attempting to analyse the structure and reproduction of this particular mode of production, we immediately encounter various theoretical problems. As opposed to his analysis of the capitalist mode of production and his brief formulations on the feudal mode, Marx nowhere constructs the concept of the Asiatic mode in terms of the theory of modes of production he develops in *Capital*. Difficulties arising from this are also compounded by the fact that comments on the Asiatic mode are scattered throughout his writings, existing at very different times in the development of his discourse; notes on its dominant form of property contained in the *German Ideology* are tainted with evolutionist notions, the accounts given of British colonial penetration of India and China (which Marx assumed to be examples of this mode of production) in the letters and articles of 1850–8[1] are extremely limited, and the overall description of the Asiatic mode given in the *Grundrisse*[2] is nothing more than an introduction to its structure, an introduction which should be read from within the theory of modes contained in *Capital*, but which in itself cannot—as we have seen—assume the status of a fully developed concept. This, then, is the problem: where Marx presents his theory, he makes hardly any mention of the Asiatic mode, since this is clearly not the object of *Capital*. Yet, in other widely disparate texts whose theoretical adequacy in analysing pre-capitalist modes is, as we have seen, open to question, he does give us descriptions of the functioning of the Asiatic mode. Consequently, these writings can provide us with *raw material* for a construction of the concept of this mode, but they cannot found the concept itself.

THE LIMITATIONS OF CONTEMPORARY ANALYSES OF THE ASIATIC MODE

It is important to stress this point, since many writers on the Asiatic mode have simply abstracted elements from these various texts and then applied them to historical data, assuming that this was equivalent to a theoretical analysis, regardless of the status or adequacy of the formulations they were operating with. Indeed, this has produced a situation in which much contemporary work on the Asiatic mode has no rigorous theoretical basis—in the form of a concept of mode of production—on which to base itself.

For example, the authors of the collection of articles contained in the C.E.R.M. document, *Sur le mode de production asiatique*,[3] operate with such a broad notion of the Asiatic mode (in which its presence can be signified simply by a centralised state having some role in the organisation of the economy and in the appropriation, distribution and exchange of the surplus produced by agricultural communities), that they are able to discover its presence in such widely differing formations as ancient Egypt, Mesopotomia, Byzantium, India, China, Persia, pre-colonial Africa and Asia, and so on. Similarly, by reading into Marx's writings in the 'Formen' his own evolutionist notion of the 'contradiction between the forces and relations of production',[4] Godelier is able to reduce the structure of the Asiatic mode to a linear development of production organised under the dominance of a kinship system, the latter simply 'evolving' into the 'centralised state' of the Asiatic mode as a result of the dominance of one family in the kinship system.[5]

We have already pointed out the limitations of Godelier's historicist discourse in relation to pre-capitalist societies, but here—as with the C.E.R.M. approach—the notion that one particular text in Marx's discourse ('The Formen') can provide us with a developed concept of the Asiatic mode of production, in isolation from the development of the discourse of historical materialism or the general theory of modes of production, leads to a situation in which the Asiatic mode becomes an empiricist model,[6] whose presence is signified by simply locating its most general aspects in a description of a particular society at a particular point in historical time. This, then, is one major error that research on the Asiatic mode has fallen into.

The other is, of course, the well known path taken by Wittfogel in various texts, notably *Oriental Despotism*.[7] It is unnecessary to reiterate Wittfogel's arguments here concerning the relation between hydraulic irrigation systems and the so-called 'Despotic State', or the characteristics of his 'totalitarian' despotic system, which is supposedly applicable to all non-basic subsistence, non-feudal and pre-capitalist societies. All I want to present is a brief critique of the Wittfogel thesis, and a short demarcation of it from Marx's concept of the Asiatic mode. This critique can be sub-divided into three aspects: first, Wittfogel's notion that irrigation works and a centralised 'despotic' state must always necessarily co-exist (since the former requires control by the latter for its very existence), has been shown to be empirically untenable, notably by Leach[8] and Polanyi.[9] Furthermore, in his text, Wittfogel himself never gives a detailed account of any one single bureaucratically administered irrigation system and its workings to substantiate his conclusions. Yet, this 'correlation' is supposedly the mainstay of his argument. Secondly, whilst Wittfogel argues that hydraulic systems give rise to centralised 'despotic' systems, he argues precisely the reverse on other occasions in the book, when he states that the land itself is transformed from an arid into an irrigated state via the action of the centralised political power. Hence, on the one hand, we have the development of agricultural production in an arid region presupposing the state and, on the other hand, we have non-arid agricultural production as a condition of existence of the state itself. Here Wittfogel's argument collapses through tautology, a contradiction that a number of writers—notably Leach—have pointed out. Thirdly, Wittfogel himself admits that despotic forms of rule need not necessarily have a hydraulic base (the device he creates to allow for such an occurrence is the ubiquitous concept of 'cultural diffusion'). If this is the case, and the 'hydraulic base' is not a necessary condition for the existence of the despotic state, then the thesis itself appears quite untenable.

Consequently, we have three major empirical and logical flaws in the thesis Wittfogel is putting forward. For these reasons, it seems to me that his notion is of limited relevance for us in our task of elaborating the concept of the Asiatic mode.

As with other appropriations outlined above, Wittfogel's thesis is similarly based on a serious misreading of Marx's notes on the Asiatic mode. In the case of his analysis—as with all other modes—Marx distinguishes the Asiatic mode of production in terms of the relations of production, and not in terms of a technique, such as the irrigation of rice. For Marx, the specificity of the Asiatic mode lies in the fact that it is a particular combination of productive forces and relations of production, not that it uses hydraulic techniques, nor that it has an 'aggro-managerial system'. Indeed, Marx would have considered such a notion technological determinism; for him particular forms of technology or a particular technical division of labour are only possible given the dominance or subsumption of the productive forces under particular relations of production. Wittfogel's conception reduces the double combination of this structure and its effects within the social formation to a tautological determinism, whose logical and empirical inconsistencies limit its explanatory value.

As opposed to these 'misappropriations', our object here will be to lay the basis for constructing the concept of the structure, reproduction and dynamics of the Asiatic mode, taking Marx's concepts in the 'Formen' and in his letters as raw material for our examination, which will be situated within the theory of modes of production as a double combination of elements as presented in *Capital*.

We will draw upon empirical raw material, largely from analyses that have been made of the Asiatic mode's existence in the Angkorian Kingdoms of Cambodia,[10] the ancient Indonesian Empires,[11] and more particularly, in sixteenth-to nineteenth-century Vietnam prior to French penetration,[12] although the bulk of this material will not be directly presented until later in the text, when we will examine the articulation of the capitalist mode within the Asiatic mode of production.

For the moment, then, our object is the structure of the Asiatic mode of production. We begin with an overall account of the major characteristics of a social formation dominated by the Asiatic mode of production, drawing on the above analyses and descriptions.

THE ASIATIC MODE: A DESCRIPTION

The organisation of production

Within the Asiatic mode, production in the labour processes is organised on the basis of the local community, the village. The unit of production is generally the family or a combination of several families. Initially, these communities are limited to the production of goods for their own consumption.

The organisation of production in the division of labour approximates fairly closely to the descriptions given by Marx of the so-called 'Indian Community',[13] or, for example, to the descriptions given by A. V. Chayanov of the nineteenth-century Russian 'natural economy'.[14] Every family unit after exchanging its goods with other family units (and even other communities) receives a value which forms the 'gross economic product' of the unit. From this is deducted a sum for material expenditures acquired during the year, and the family is then left with the value that it has produced for itself for its own consumption over the course of the year. This sum then becomes the basis on which production for the coming year is calculated. The important point here is that the family unit's 'income' *has its own determinants*—the size and composition of the family, the productivity of the family working unit and the degree of self-exploitation undertaken by family members. The latter, which is absolutely crucial for planning the output of the family (given that the other two determinants are not easily alterable), again has its own causes. To quote Chayanov on this point:

> The degree of self-exploitation is determined by a particular equilibrium between family demand for satisfaction and the drudgery (irksomeness, laboriousness) of labour itself. Each new monetary unit of the growing family labour product can be regarded from two angles; first, from its significance for consumption for the satiation of family needs; secondly, from the point of view of the drudgery that earned it. It is obvious that with the increase in produce obtained by hard work, the subjective evaluation of each newly gained monetary unit's significance for consumption decreases; but the drudgery of working for it, which will demand an ever greater amount of self-exploitation, will increase. As long as the equilibrium is not reached between the two elements being evaluated (i.e. the drudgery of the work is subjectively estimated as lower than the significance of the needs for whose satisfaction the labour is endured), the family, working without paid labour, has every cause to continue its economic activity. As soon as this equilibrium point is reached, further labour expenditure becomes harder for the peasant or artisan to endure than is foregoing its economic effects.[15]

The object of production in the family unit is to reach this 'equilibrium point', where the family's needs can be met and the 'drudgery' of labour is not subjectively evaluated as excessive in relation to attaining these needs. Once this point has been achieved, further production becomes, as Chayanov states, 'pointless', and the family utilises natural conditions, its own composition (the changing ratio of producers to consumers over the course of a generation), exchange, and so on, to achieve 'equilibrium'. What this means is that tendencies to increase output through capital accumulation and exploitation of new agricultural land are limited by the composition of the family labour force and the drudgery of work in relation to consumer needs. If the subjective equilibrium point can be reached by utilising the family labour force, then there is simply no need to increase drudgery by cultivating new land or to improve productivity by using present income to purchase more efficient tools. Conversely, if the equilibrium point cannot be attained (as, for example, when the husband and wife are working to support young children who necessarily are not productive), then, in the short term only, the above measures will be taken, only to be set aside once equilibrium is attained. Thus, the use of income to purchase new tools or the cultivation of new lands will only be acceptable to the family unit if, in the short term, it can ultimately produce a new equilibrium with less drudgery in labour expenditure and greater satisfaction in family consumption; as soon as this point is reached, such measures can be abandoned.

In this labour process there is, therefore, a constant tendency to reproduce the conditions necessary for family consumption—such is the object of production.

What of non-agricultural activities in this division of labour? The little artisan industry that does exist is dependent either on the needs of this production or on the immediate requirements of the community. Initially, artisan work in the Asiatic mode does appear to have been confined to periods when labour-time did not have to be totally used up in agricultural production. In this, as in agricultural production, the community is initially organised around the production of use-values for immediate consumption.

With the development of the Asiatic mode of production, this organisation persists, but it is increasingly supplemented by the production of commodities for exchange, either with other communities, the state, or (via the state) with other social formations. In this way, it seems that, with the development of the Asiatic mode, the original agricultural and artisanal 'autarchy' of the communities begins to be overcome. This is strikingly seen, for example, in the seventeenth-century Vietnamese kingdoms of Nguyen and Trinh, where, under a state monopoly, production of such commodities as wood products (particularly paper), metal products, and textiles were organised both for exchange with other areas (Japan, China), and for internal circulation amongst the various communities themselves.[16] This process of supplementing production of exchange-values for production of use-values is generally accompanied by a monetarisation of the economy, and, of course, by the state's importation of commodities—for the use of its own functionaries, for its organisation of the means of production or for its ideological requirements (see below).

The important point to note here is that, although this process of production of exchange-values is a general tendency within Asiatic modes, it is one that is always *restricted within the confines of boundaries laid down by the state itself.* In order to examine this point further, we need to analyse state intervention within the process of production.

The intervention of the state in the process of production

As with other modes (such as the feudal), the state appropriates surplus labour from the communities, usually in the form of a rent-in-kind, which necessitates the application of labour above and beyond that required for local consumption and exchange. For example, in the Vietnamese kingdoms we have mentioned above, there existed direct taxes levied by the state in the form of a land tax on communal land, a head tax, a tax levied for use in state ceremonials and buildings, and indirect taxes on the produce of the soil, the circulation of commodities and transport. *However, as distinct from other modes, the state also plays a crucial role in the reproduction of the prerequisites for production itself; in this way, as we shall see, its role in the process of production contributes to the reproduction of the Asiatic mode itself.*

Whether it be by organising the surplus-labour time of armies of communal labour from the villages to construct the irrigation works (dykes, canals), essential for village agricultural production to take place along the river valleys (for example the Vietnamese and Angkorian kingdoms),[17] or by regularly redistributing communal lands to meet the changing demographic requirements of the different villages (some holding a surplus population which existing village production is unable to feed, whilst others are producing surpluses beyond what is needed for them by both village and state—for example the Vietnamese kingdoms in the eighteenth century),[18] or by organising the storing of produce to meet requirements during the dry seasons,[19] or by organising the rotation of crops,[20] or the production and distribution of raw materials for agricultural production (mining, metallurgical industries, etc.), the state fulfils a crucial economic function in the Asiatic mode by its intervention in the village economies. *It plays a part in organising the means of production.* In this way, it directly contributes to the process of production itself. As a result, the reproduction of village production and consumption is assisted by the organising function of the state.

State ownership

The state extracts surplus labour from the communities in the form of collective labour performed as a means of production (as above). In addition, it also extracts surplus in the form of a tribute paid to the state, existing in one or several of the taxes indicated above. The determinate relation of production which governs the extraction of surplus labour in the Asiatic mode is, therefore, one existing between village and state; it is this relation of production which (as we shall see) structures the labour processes operating in the division of labour in the villages. The state's extraction of surplus labour and its organisation of production are expressed through an ideologically-defined ownership of the territory within which production occurs, and a subsequent control over the allocation of property to the villages; in each case, the state both owns and controls the land, and is ideologically defined as the owner of the soil.[21] Consequently, the state has the right to restrict any development of private property in land, and—as a result—to prevent the accumulations of capital obtained either from trading or usury from being invested in private ownership of the soil. This is seen, for example, in the Vietnamese kingdom of Trinh in the eighteenth century, where, in addition to the traditional gifts of land to officials by the Emperor, state functionaries (mandarins) increasingly began to appropriate land through their control of the process of village land redivision. This, however, was held in check by royal decrees which made ownership of land non-hereditary, the 'private' ownership returning to the state on the death of the individual functionary. Furthermore, as Than Nha points out in his analysis of this period, the granting of land itself was always ultimately regulated by the prevailing subsistence population ratio, and the state's attempts to balance this.[22]

The control over state power

Thus far, we have indicated several features specific to the Asiatic mode but, up until now, we have spoken of 'the state', and before going any further we should be more specific.

State organisation is generally subdivided into: (1) a military organisation, comprising a vast standing army recruited from the villages to defend the state against rival kingdoms; (2) an administrative organisation, governing the regions and responsible for employment, recruitment and training of functionaries, the levying of taxes, state finances, state ceremonies and 'justice'; (3) the organisation of public works and the running of state industry and the organisation of internal and external trade.

To examine the second point—who holds power within this apparatus? Whether we examine the Empires and kingdoms of Angkor, Indonesia or Vietnam, we find that state power devolves on a lineage and family, and—ultimately—on an individual monarch, whose rule is legitimised through birth, genealogy and tradition. The actual apparatus of the state, whether its administration is centralised or decentralised, may be sub-divided into departments controlled by a whole series of functionaries, carrying out ceremonial, administrative, religious, intellectual and other functions, but at the head of the apparatus is always the reigning monarch and his lineage or family. The ideologies within which the political power of the monarchy is expressed and the general conception of the relation between king, state and communities in a social formation dominated by the Asiatic mode contains elements that recur repeatedly. Here I can only briefly summarise the vast amounts of research that have been carried out on this subject.[23]

The concept of kingship is such that, generally, there is a unity postulated between the universe (the 'cosmos'), a divinity(ies), the king (who himself possesses divine status), the political, legal, and economic functions that the latter fulfils in relation to the village communities, the state (whose functions are 'ordained' by the cosmos), the process of production, and the fertility of the soil itself. The task of the monarch—instructed by the divinity(ies)—is to ensure that the communities are economically prosperous. This he does by organising such means as irrigation systems and public works, through his knowledge of the seasonal cycles governing agricultural production, by his organisation of stores, exchange of commodities, and so on. Through these actions, he guarantees agricultural fertility, and, 'in return' his subjects pay him tribute in the form of service, taxes, etc. The monarch is constantly guided by the deity, and, consequently, his tasks are always expressed through religious symbols and rituals, this necessitating the utilisation of appropriated surplus in the construction of temples and other symbolic buildings, as well as the maintenance of a body of religious functionaries to guarantee the success of his endeavour. As a result, the practice of the monarch is totally fetishised, the actual relations existing between the monarch and the village communities, and, indeed, between the state and these communities, appearing in an inverted form in the ideology of kingship. Thus, what appear ideologically as the divine attributes of the king (e.g. the organisation of public works, the redistribution of land) are merely effects of his function in the appropriation, distrubution and utilisation of surplus labour, either in the form of labour itself, or in the form of commodities. The king, ideologically perceived as the 'guarantor' of fertility, can only be such given the dominance of the exploitative relations existing between state and village. Similarly, the ritual which guarantees the 'correct' practice of the monarch can only exist as an effect of the distribution of realised tribute to the state, and so on. This fetishism can be seen to reach its highest form when, for example, the failure of the harvest, or the low returns received from exchange, are directly attributed to the inadequacies of the monarch, to his 'incorrect' application of ritual. To these specific features of the Asiatic mode—its form of surplus extraction, the essential intervention of the state in the labour process, the state's organisational role devolving from its absolute control over the distribution of the surplus realised, its particular form of 'property', and its dominant ideology—we must now add one more which, as we shall see, is important for discussing the dynamics of the Asiatic mode.

The 'unchanging' basis of the Asiatic mode of production

As we have tried to show, the villages are essentially entities that can survive under the dominance of different forms of Asiatic state power; no matter which lineage, family or monarch holds power and controls the state apparatus, the communities will survive, assisted by the state exercising the economic functions we have outlined above in relation to the villages. Thus, it seems that, whilst a social formation dominated by the Asiatic mode can be politically highly unstable (one dynasty or family replacing another in the state apparatus), the overall structure of the social formation can remain extremely stable. New political systems can constantly be constructed on the economic basis of the villages, but this has little or no effect on the reproduction of the mode itself. Hence, the histories of such social formations as Vietnam, Cambodia and the Indonesian Empires are politically turbulent 'on the surface'[24] yet their overall structure remained essentially unchanged for centuries, right up to French or Dutch colonial penetration.

This, then, is a further characteristic of Asiatic modes—their changing yet 'unchanging' history, a phenomenon that results from successive political rulers occupying state power, yet retaining intact the functions of the previous ruler and the economic relations with the villages. Only colonialism and imperialist penetration were, in fact, as we shall see, able to break down this cycle.

Having briefly described the organisation of a social formation dominated by the Asiatic mode, we must now turn to a short theorisation of its structure, reproduction and dynamics from within the theory of the elements and their combinations, as presented in *Capital*.

A THEORISATION OF THE ASIATIC MODE

In the Asiatic mode, the direct producers are in possession of their own means of subsistence and means of production, no matter what the different labour processes are that they produce in (agricultural labour for immediate consumption or exchange, collective labour on communal works, labour in mineral extractive industries, individual artisan labour or work in the state-owned units of 'industrial', production, etc.). Under these conditions, as Marx states, surplus labour can only be extracted 'by other than economic pressure'. Consequently, in order that the state can extract surplus labour in the form of tribute, in order that it can own and control the distribution of the surplus product, the intervention of another level of the social formation within the economic is required. In the Asiatic mode, this level is primarily the ideological, since it is, above all else, within ideology that the right of the state to the surplus labour produced is defined and legitimised. The state, whose apparatus is controlled by the monarch and his family, exercises its right to the surplus via genealogy and tradition, usually expressed, as we have seen, in religious and philosophical ideologies, embodied in particular rituals and practices exercised through the political power of the state. Whilst the monarch requires the state apparatus to realise his decisions, his power and rights over the system of production are legitimised ideologically (as opposed, for example, to the feudal mode, where the noble's right to surplus in the form of ground-rent is achieved through a political intervention). Surplus labour is extracted because the monarch has an 'ideological right' to it, a right accepted by the communities and expressed in religion, art, literature, and so on, throughout the social formation. It is this, of course, which produces the very fetishism of the ideological instance within the Asiatic mode that Marx repeatedly refers to:

> Since the *unity* is the real owner . . . it is perfectly possible for it to appear as a particular being above the numerous, real, particular communities. The individual is then in fact propertyless, or property . . . appears to be mediated by means of a grant from the total unity to the individual through the mediation of the particular

> community. The despot here appears as the father of all the numerous lesser communities, thus realising the common unity of all. It therefore follows that the surplus product . . . belongs of itself to this higher unity.[25]

Yet, the role of the state does not end with the appropriation and distribution of surplus labour. As we have seen, it also acts as *an organiser of the means of production* through its control over the use of communal labour applied to the process of production. Consequently, the state's dominance through its extraction of surplus labour is, as it were, reinforced by its organising role in the labour process. What this means is that not only can the Asiatic mode itself not be reproduced without the reproduction of its particular determinate relation of production expressed in an inverted ideological form, but that production in the labour processes is facilitated by the intervention of the state in the organisation of the means of production. Thus, the reproduction of production itself is, in this way, dependent upon the organising role of the state in its functions of land redistribution, irrigation, etc. This role reinforces the fetishism of the state and the ideology within which political power is exercised. The effects of communal labour, the *sine qua non* of the state's ability to organise its public works, thus appears as the work of the state itself:

> The communal conditions for real appropriation through labour, such as irrigation systems (very important among the Asian peoples), means of communication, etc., will then appear as the work of the higher unity—the despotic government which is poised above the lesser communities.[26]

Thus, we can conclude that the structure of the Asiatic mode determines a particular articulation of the structure in which, in order for the surplus labour to be extracted from the direct labourers, and for state-power to intervene in the process of production, the ideological instance must be determinant. It is the relations of production existing between the state and the village communities that ensure the state's control over the production of surplus labour, and the major prerequisite for this exploitative relation is ideological, even if it is necessarily exercised through the use of the state apparatus. Within the Asiatic mode, then, *the reproduction of the determinant relation of production is ultimately ensured through the determinancy of the ideological instance*.

The *dynamics* of this mode result directly from this determinancy of the ideological and the consequent power of the state. As opposed to other modes of production, such as the feudal or capitalist, the dynamics of the Asiatic mode establish no real possibility for laying a basis for a transition to the dominance of a new mode of production. When Marx wrote that the Asiatic mode superficially appeared to be 'unchanging', or when he spoke of its 'stagnation' (without a history) he was not, as a number of writers have thought, simply 'ignorant of Asian social history',[27] or 'Europocentric',[28] but he was making a crucial theoretical observation about its dynamics—namely that they result in a particular form of non-development which affects the overall structure of the Asiatic social formation. Namely, that the determinancy of the ideological, embodied in the actions of the state, places severe restrictions on the processes of capital accumulation, the private ownership of land or capital, and, of course, on any separation of direct producers from their means of production (except for occasional seasonal or cyclical labour in the state 'factories' or mines). By its total control over these processes, the state prevents these possibilities—any of which could establish the possibility for a basis from which a transition to another mode of production could begin to develop—from being reproduced. As a result, the action of the state in its role as the owner of land and capital, which devolves on it from its place in the reproduction of the Asiatic mode, prevents the development of any transitional tendencies. The development of private ownership in land, the constant use of direct labour for purposes other than village production (e.g. labourers organised collectively under the dominance of an artisan), or the accumulation and realisation of capital in either of these processes would pose a serious threat both to the general organisation of village production and the economic relations between villages required by the Asiatic mode, and to the economic functions and basis for the power of the state. These factors alone, apart from the pervasive ideological barriers to such developments, are sufficient to prevent their realisation. Consequently, within the Asiatic mode, as long as the ideological level retains its dominance, and as long as the determinate relation between state and village communities is reproduced, the development of transitional tendencies will always be severely restricted, and, so, therefore, will be the development of the productive forces themselves.

This restriction of transitional tendencies is, of course, maintained in the histories of social formations dominated by the Asiatic mode, despite the very 'dramatic' changes that characterise their political history. State power, as we have noted, is repeatedly seized by one dynasty, only to be defeated and replaced by another invading dynasty, in a highly discontinuous process. Yet, despite this political discontinuity, the relations between village and state that we have examined above are continually perpetuated. No matter what dynasty occupies state power, the functions of the state in relation to the village continue to be reproduced. The determinancy of the ideological in the reproduction of the Asiatic mode is constantly secured, and, in this way, the dynamics peculiar to this mode's non-development are continually realised. Hence the history of these formations is perpetually changing yet essentially unchanging—a paradox resulting from the specificity of the dynamics of the dominant mode of production.

Such, then, is the structure, reproduction and dynamics of the Asiatic mode—a mode dominant in a number of S.E. Asian social formations prior to capitalist penetration in its different forms.

In this section, we have tried to show how the concepts of historical materialism can be used to analyse a particular non-capitalist mode of production. Other modes could have been constructed, based on research from other areas, but this would have been beyond the scope of our object here, which is simply to *indicate how these concepts can be used in analysing modes dominant in pre-colonial, non-capitalist societies*.

If we were analysing the history of non-capitalist societies prior to capitalist penetration, we could now take this framework as our theoretical basis. Our object, however, also includes the specification of a theoretical foundation for analysing the effects of capitalist penetration in its various forms on non-capitalist formations. Consequently, we now need to outline how we can theorise the articulation of the capitalist mode in its different stages of development within the dynamics of non-capitalist modes of production.

We have already noted the existence of these stages and their effects, but to approach this question with any rigour, we must analyse the meeting of two processes—the reproduction of the non-capitalist mode of production and the effects within this of each stage of capitalist penetration. We have specified how the problem of reproduction and dynamics can be approached in relation to a particular non-capitalist mode of production, and indicated how the determinants of the different forms of penetration—under merchants' capital, competitive capitalism and imperialism—can be theorised from within the dynamics of the capitalist mode of production. We now come to what is perhaps the most important point. What are the major effects of the meeting of these processes on the non-capitalist mode of production? Only by attempting to answer this question will we be able to analyse how the elements of a specifically capitalist mode of production come to be reproduced within a non-capitalist social formation, and eventually, how a developed capitalist mode comes to reproduce itself and become dominant over the non-capitalist mode. Once we have established a theoretical framework for analysing this process of articulation of capitalist with non-capitalist modes, we then have a basis for examining the way in which it determines in the last instance the class structure, the state, ideology, and so on. Before entering such an analysis, we must, however, complete our examination of the determinants of this articulation of modes of production in the Third World formation by analysing how the effects of capitalist penetration in its different forms can be theorised.

10 The Effects of Capitalist Penetration of Non-Capitalist Modes of Production: Penetration Under the Dominance of Merchants' Capital

In earlier chapters we noted the role that accumulations of mercantile capital played in the genealogy of the capitalist mode of production, and that the dominance of this form of capital in the period of the dissolution of the feudal mode and the transition to dominance by the capitalist mode had specific economic effects on non-capital modes.[1] We also indicated that these effects continue to be reproduced even when merchants' capital is no longer dominant, when penetration occurs under other forms, as the circulation of mercantile capital is increasingly subjected to the requirements of capitalist production.[2]

In this chapter, we will briefly examine the *major effects of merchants' capital*, since the agrarian heritage that it left, particularly in Latin America and parts of Asia, persists as a barrier to capitalist development in these countries, and it is essential that we formulate a theoretical basis for analysing this phenomenon.

The major economic effect of penetration under merchants' capital is the reinforcement of already existing forms of extra-economic coercion in agricultural production in the non-capitalist mode of production. Merchants' capital achieved this either by creating forms of landed property and relations of production similar to those dominant during the European feudal period, or by simply utilising and then perpetuating existing relations of production. Examples of the former are Spanish, Portuguese, and (to a much lesser extent) early English colonialism, and, in the latter case, we have the example of Dutch colonialism, which, as

Engels aptly stated, 'organised all production on the basis of the old communistic village communities.'[3]

Having made this assertion, we will now try to validate it by posing the following question: to what degree *did* penetration under the dominance of merchants' capital assume this form? What, briefly, were the major characteristics of the variants of this form? What are the major consequences that the reproduction of these variants have had on the development of the non-capitalist mode after the decline of merchants' capital and the emergence of new forms of penetration?

EXAMPLE I: THE CASE OF SPANISH AND PORTUGUESE PENETRATION OF LATIN AMERICAN FORMATIONS

To begin to answer the first question, it is essential briefly to consider the example of Spanish and Portuguese penetration of Latin American formations from the early sixteenth century.

As is well known, the relatively early and extensive development of mercantile trade in seigneurial Spain and Portugal was accompanied by an extremely late (twentieth-century) capitalist development. During this period, mercantile capital, largely under state control (particularly after the seventeenth century) remained the dominant economic force, and—for this reason—its effects on its colonies and dependencies in Latin America were extremely pervasive. Consequently, they provide us with an excellent example of the effects of mercantile penetration.

Beginning with a relatively simple form of plunder ($M - C - M^1$), in which private individuals with their armed coteries rounded up gangs of indigenous labourers to work as slaves in gold and silver mines, exporting their surplus product to Iberian or European markets, Spanish and Portuguese colonial penetration transformed itself, in the seventeenth century, into a qualitatively new form, to produce what came to be known as the encomienda. These were initially lands appropriated from local chieftains by the Spanish or Portuguese State, and then given to individuals as gifts in compensation for their services in the Latin American conquests.

In many ways, the 'encomienda' were organisationally equivalent to the 'manor' of European feudalism. The owner of the land was legally entitled to a portion of the peasantry's surplus product (in the form of a corvée or more general tribute), part of which was utilised to feed the labourers in the mines, and part of which was exported to Europe. In some areas, the peasantry retained their own landholdings, and worked for a part of their labour-time on the encomienda, whilst in others, where the soil was less fertile or the population relatively less dense, slavery was prevalent. The most extreme form of the latter developed of course, in the Portuguese colonies, where slave labour began to be imported to encomiendas from Africa in the sixteenth century.

Consequently, it appears from its initial stages that this form of penetration was characterised by a transformation of the various systems of forced labour that pre-existed it. Spanish and Portuguese colonialism replaced the previously dominant process of extraction of surplus labour—in which local chieftains or divine monarchs could appropriate and distribute part of the surplus product produced by the peasantry in a variety of different modes—by forcing direct labourers to produce within an agrarian division of labour structured by relations of production that rapidly came to assume a feudal form.

This process was further intensified from the middle of the seventeenth century onwards for a number of reasons, which can be briefly specified. The encomienda had always held a secondary position relative to the gold and silver mines, which provided most of the surplus appropriated by the state. When the production of these minerals began to decline in the seventeenth century, the encomienda gradually lost both their internal markets and the means of transporting their products to Europe. In fact, at the beginning of the eighteenth century, the Spanish state officially abolished the encomienda. This official abolition, however, only had the effect of strengthening the landowning colons' internal control over the direct labourers. As Furtado points out:

> In effect, once the *encomienda* system had been abolished, it was the control of land that made it possible to continue extracting a surplus from the native population. Since this surplus, by its very nature, had to be used almost entirely locally, the social structure tended to assume the form of isolated or semi-isolated communities. . . . The ownership of land because the basis of a system of social domination of the mass of the people by a small ethnically and culturally differentiated minority.[4]

As a result of this developing process of isolation, accompanied by the desperate attempts of ex-encomendieros to diversify their production by concentrating on various crops for export throughout the eighteenth century, the Latin-American plantation (*hacienda*) came to assume the form by which we now know it.

Created by mercantile penetration, the various aspects of its coercive structure were intensified by its continued adherence to export agriculture. The expansion of the external market in the eighteenth and nineteenth centuries thus consolidated the hacienda system rather than acting as a disintegrating factor. Furthermore, it mattered little whether or not the colony broke away from Iberian political dominance (as was the case with Brazil, Chile), or whether it remained under its control (e.g. the Phillipines, Cuba), since the agrarian structure developed in much the same way, the only major exception being that the surplus product obtained from exporting cash-crops was appropriated largely by the state rather than by the individual plantation-owner. Consequently, the major effect of mercantile penetration, with its object of accumulating a surplus through inter-continental trade, was intensified through increasing production for export. The structures that had been created by merchants' capital were strengthened by the perpetuation of mercantile policies throughout the eighteenth and nineteenth centuries.

The plantation was not, of course, simply an economic form. Whether it developed under the tutelage of Iberian colonialism, or, as in the case of the West Indies, under the dominance of British mercantile capital, the plantation produced ideological and political forms and institutions that were specific to it. These have been fully analysed elsewhere,[5] but we can briefly mention them here, since they are important for our subsequent analysis of the consequences of mercantile penetration for the structure of Third World formations during the post-colonial period.

As a result of the hacienda system, a small class of landowners came to own and control the most fertile agricultural areas. By applying a massive, low-paid, labour-force on their landholdings, this class was able to reap tremendous profits from the export of cash-crops; it became possible to expropriate a relatively large surplus product from a labour force with extremely low productivity. Those without the capital necessary to produce land had two choices—either to work as a labourer on the hacienda, or to try to cultivate the relatively less fertile agricultural land untouched by the plantation owners. Consequently, alongside the hacienda system there developed a variety of systems of land tenure. For example: workers spending most of their available labour time on the plantation and devoting the remainder to subsistence cropping on their small plots, sharecropping, growing crops on their own land and sending them to the hacienda as a form of rent in kind, and so on. The output of these peasant plots was, of course, pathetically small when compared with the plantation and its vast resources. As a result, the agrarian structure of Latin America came to be dominated by the hacienda, which also became a basic feature of the social structure.

The dominance of the hacienda produced a series of isolated, and rigidly stratified units, dominated by what can only be termed an extremely paternalistic ideology, in which the landowner's aim of the maximum possible exploitation of his labour-force was articulated in contradictory notions that equated the hacienda with extended familial relations, yet stressed the inherent inferiority and submission of all those who performed manual labour. The hacienda was structured on both caste and class lines, the dominant relations of production and social relations existing between landowners, administrators, overseers, artisans and workers being overlain by racial cleavages between immigrant and non-immigrant labour. Generally, the dominant position of the landholders was legitimised through an extremely pervasive religious ideology, initially based on Iberian catholicism, but later skilfully blended with local cults, in which their economic and political positions were immutably eternalised.[6] Within the isolated hacienda, political power and control devolved solely on the landowner and his family, assisted by hired militia on occasions when the heavy hand of ideology occasionally failed to produce the desired response from the labour force. Consequently, the unification of the country, either under Iberian colonialism or during the independence period, was carried out precisely by those who held political power at that time, namely the hacienda-owners. State power came to rest totally on alliances between plantations and regions. Until the emergence of commercial groups, the industrial and comprador bourgeoisie, or the rural and urban proletariat in the late nineteenth and early twentieth centuries, the political dominance of the landowning class in Latin America was absolute.

These various aspects of hacienda organisations, set up as effects of Iberian mercantile penetration, and intensified by the production of export crops, were continually reproduced within the social formation of Latin America throughout the eighteenth to twentieth centuries, and their feudal economic and political forms were increasingly forced to exist alongside a developing capitalist sector whose penetration gradually threatened their economic dominance in the agrarian sector. Yet, this capitalist sector, in its initial stages, was forced into a political alliance with the landowners in order to secure the preconditions for its own growth and development. This has produced (and still produces) a very contradictory situation in the recent history of many Latin American societies, since the articulation of the two major modes of production—a feudal and a capitalist mode—has produced changing and vacillating political alliances between the various classes, the emergence of a variety of strategies for developing the agrarian sector,

and has transformed the ideologies within which much of the rural population live, their relations to their subsistence world and their enforced submission to the dictates of their landowning overseers.

We can see, therefore, that the whole question of the effects of this articulation of a feudal and capitalist mode of production, and the structuring of its development by imperialist penetration is crucial for any contemporary analysis of Latin American societies. We will return to the question of this articulation below. What we want to stress here is that the basis for the reinforcement of coercive relations and the emergence of feudal forms of production was an effect of one specific form of capitalist penetration, namely that dominated by merchants' capital.

EXAMPLE 2: DUTCH COLONIALISM AND THE INTENSIFICATION OF EXISTING FORMS OF EXPLOITATION

If we now turn to another example of mercantile penetration, it becomes clear that, rather than developing feudal or quasi-feudal forms, merchants' capital can also achieve its result simple by an *intensification of the forms of exploitation already in existence in the non-capitalist formation*. This is the case, for example, with Dutch colonial penetration in S.E. Asia, and particularly in Indonesia.

Beginning in the early seventeenth century, Dutch mercantile penetration inserted itself into the East Indies trade, and gradually secured a monopoly, excluding other European powers. By the end of the eighteenth century, it was in a strong enough position to force Javanese peasants to produce crops for European consumption, and, despite a brief period of British rule at the beginning of the nineteenth century, a system of forced cultivation was firmly established, under state control, by the 1840s.

Under this 'culture system', peasants were forced to grow commercial crops (sugar) on part of their lands, or, alternatively, to devote part of their labour-time to the cultivation of government crops (coffee) on local wasteland. The entire system of production and the immense commercial profits that were realised from it on the European market was based on a simple process of intensification of the existing economic system. The surplus labour previously extracted from the peasantry by indigenous ruling groups (the Hindu *pryayi*) and trading groups in the form of corvée, tribute or commodities, was now redistributed under the supervision of Dutch officials, and the existing forms of exploitation were massively intensified. As a result of their being forced to cultivate export crops, the peasantry now had less labour time and less land to devote to their own susbsistence, and fewer agricultural products to exchange for the goods they could not produce. Consequently, a peasant economy that had largely been self-sufficient was rapidly destroyed by the culture system, and landlessness, indebtedness and hunger emerged to dominate the countryside.[7] The effects on the agrarian organisation of Java (containing 85 per cent of Indonesia's population) were extremely profound. This restructuring of production enforced by Dutch colonialism still dominates the agrarian sector, which continues to rely massively on the production of cash crops for export, and is consequently characterised by the existence of minuscule plots and landlessness.[8]

The main point to be stressed here is that from the moment Dutch mercantile capital began to go beyond its strictly mercantile functions of exchange—purchasing commodities in the Indonesian archipelago to sell at high rates of profit in Europe—from the moment when, in Marx's phrase, it began to 'rule over production', it intensified pre-existing non-capitalist, servile economic relationships. In so doing, it necessarily reinforced the barriers to any future process of separation of direct producers from their means of production, to any process that attempted to develop the agricultural sector in a capitalist direction.

We could quote other examples: Portuguese mercantile penetration of W. Africa, where the dominant relations of the lineage mode of production were reinforced by the colonial slave trade, intensifying the power of the lineage elders by enabling them to accumulate more 'élite goods' through their key role in the exchange of slaves. These, however, would only reinforce our conclusion: that penetration of non-capitalist modes under the dominance of merchants' capital appears, in all cases, either to have created economic forms similar to those existing in feudal Europe, or to have reinforced pre-existing relations of production. In either case, the unity of direct producers with their means of production was strengthened, as opposed to later forms of penetration, which, as we shall see, tended to break down this unity and, in so doing, established a basis for the development of capitalist production.

THE BASIS FOR THE EXISTENCE OF THESE EFFECTS

Having reached this conclusion, we should now briefly look at the question of why mercantile penetration produced these specific effects.

The existence and continuing reproduction of the encomienda and then the hacienda in Latin America has to be examined in relation to the contradictory relations that they experienced with social formations which, during the sixteenth to nineteenth centuries, began to enter into a transition which became effectively blocked or retarded even up to the twentieth century itself.

During the late sixteenth century, both Spain and Portugal were essentially feudal formations. The political power of the monarchy and seigneurs was still largely unchallenged, and the recent decline in the numbers of the peasantry had been overcome by importing slaves and enslaving immigrant labour. Consequently, given the availability of land for military conquest and the possibility of utilising large supplies of labour (both internally and from Africa), it is hardly surprising that the encomienderos should attempt to reproduce agricultural units similar to those existing in their home countries during the colonial period. The prerequisites for this existed in the colonies, and no other possibility had actually presented itself.

To explain the intensification of this form in the mid- and late seventeenth century is, however, rather more difficult. With regard to those colonies that attained their independence in the early nineteenth century, it seems that the decline of the gold and silver trade that we mentioned earlier, together with the resultant decline of the encomienda, led landowners to examine the possibility of diversification for export. This coincided, in the late seventeenth and early eighteenth centuries, with an expansion of merchants' capital in several European countries, notably England and France. This expansion—although this remains a contentious point[9]—seems to be the result of the limitations placed on the internal investment of accumulated merchants' capital by the very slow development that embryonic capitalist production experienced during the period 1620–1750. This slow development—as the economic historian, Lublinskaya, has pointed out[10]—resulted largely from the difficulties that capitalist production experienced in transcending the limitations imposed by a technical division of labour still based on the forms of organisation inherited from artisan production. Capitalist production also had difficulties in developing a home market, since its failure technologically to advance significantly the forces of production prevented it from destroying rival peasant artisan manufacture in the countryside. In addition, the power of the craft guilds[11] combined with the persistence of artisan production, restricted the flow of a labour force from the countryside to the urban areas. All these factors continually retarded capitalist development during this period, and it seems that, in this situation, both colonial markets and colonial trading increased fairly rapidly, whilst the supply of export commodities, particularly from Latin American countries, began to develop.

We thus have a conjunction of events that were particularly favourable to the development of the hacienda. Its limited technical requirements could be met by European exports, and the trade in the commodities that it produced and exported also increased. Consequently, the situation that characterised the transition from feudalism to capitalism in European countries such as England and France was particularly favourable to the development of the hacienda at the very period when it was establishing its dominance in the agrarian structure of Latin America.

In addition to this, as a result of their capitalist tendencies being increasingly blocked, the Iberian countries became ever more dependent on their trading in Latin American export crops towards the end of the eighteenth century—and still so during the nineteenth. In Spain, the feudal landowning class constantly curbed the emerging capitalist class by its absolute control over the state; although the influx of American gold and silver had stimulated the development of small textile, silk, leather, and iron works, these were given no state assistance and remained stillborn due to a vigorous guild system, heavy taxes, an appalling internal transport system, and an inadequate supply of available labour-power.[12] Consequently, the profits obtained from colonial trading were either invested outside the country, or were used to strengthen the political dominance of the feudal regime.

In Portugal, the results were similar. Slavery here was actually perpetuated throughout the seventeenth and early eighteenth century, and the peasantry were subjected to increased feudal exploitation. Rather than developing any mechanisms that encouraged the separation of direct producers in the agrarian sector, Portugal specialised in the re-export of commodities and capital obtained from its colonies to other European markets, and, in return for the former, it received goods purchased by the landed nobility, which further strengthened the latter's power. The alliances between the state, the landed nobility, and merchant capitalists remained constant during this period, and, as a result, factors similar to those operative in the Spanish Empire prevented the emergence of capitalist production and strengthened mercantile interests. These developments in the Iberian peninsula reinforced the position of the hacienda, whether it produced under the control of Iberian colonialism, or simply benefited from either Spain or Portugal continuing to trade in its products.

For these reasons, then, it seems that the hacienda was strengthened during the seventeenth and eighteenth centuries by the continuing dominance of merchants' capital in the European countries, a dominance that can only be understood by situating it in the varied capitalist genealogies experienced by different European countries during this period, whether they were blocked (as in the case of Spain and Portugal), or retarded (in the case of England, France and other countries).

Similarly, when we examine the reasons for the continued dominance of merchants' capital from the seventeenth right through to the nineteenth century in Holland, the reasons are fairly clear. As Hobsbawm notes,[13] the co-existence of two modes of production can be clearly documented during this period—an embryonic capitalist production emerging from within the feudal order during the eighteenth century. Here again, though, limits were imposed on the development of Dutch capitalism by the political dominance of the seigneurial classes in the landward provinces and by the policies of the craft guilds, which were still extremely strong even in the most economically advanced towns in the eighteenth century. As a result, the state remained totally committed to policies of foreign investment and overseas commerce, even up to the mid- and late nineteenth century. The effects of this situation on the Dutch colonies were striking. The transition from a strictly trading role to one of rigorous control over colonial production was made at a relatively late stage, by the late eighteenth and early nineteenth centuries, and from this period onwards, Dutch colonialism increasingly intensified its exploitation of its colonies, reinforcing non-capitalist agrarian relations in an attempt to salvage its declining economic position *vis-à-vis* the growing strength of the other industrialising powers. The rampant coercion of the culture-system in Java amply testifies to this, as do the transformations in this policy carried out during the 1860s–80s, when the gradual transfer of plantation ownership from the state to individuals and corporations actually intensified the dominance of the plantation and the reliance of the economy as a whole on the export of agricultural crops.

Thus the intensification of non-capitalist relations was essentially an effect which must, again, be initially situated within the structure and reproduction of the Dutch economy and, in particular, in the continuing dominance of merchant's capital throughout the colonial period, due to the relatively late emergence of capitalist production.

MERCANTILE CAPITALIST PENETRATION AS A BARRIER TO CAPITALIST DEVELOPMENT

From these brief examples, we can see that, in whatever form, penetration under the dominance of mercantile capital tends to intensify the pre-existing unity between direct producers and their means of production. Whether the peasant works for the land-owner as a share-cropper, or whether he cultivates his own subsistence plot in addition to labouring on the hacienda, or whether the production on his own land is regulated by the landowner—no matter what system of land tenure prevails the peasant remains in possession of his means of production, tied to the soil. What distinguishes penetration under merchants' capital from other forms, therefore, is that *whereas* (as we shall see), *the latter assist and then create the basis for capitalist development in non-capitalist modes of production, the former produces the dominance of relations of production that will later act as a barrier to capitalist development.*

As we shall see later,[14] imperialist penetration was able to break down the reproduction of non-capitalist modes of production, and forcibly separate producers from their means of production. Yet, in social formations previously penetrated by merchants' capital, this process was, in general, more difficult, *for reasons directly related to the effects of this penetration.*

First, because the resultant political power of the landowning classes enabled them to restrict penetration by indigenous capitalist groups and foreign capital in the agrarian sector.

Secondly, because the reaction of imperialist states to the non-capitalist production set up by merchants' capital has remained contradictory since the beginning of the increase in the export of capital to Latin America in the late nineteenth century. Whilst on the one hand requiring the retention of the non-capitalist agricultural sector in order to secure cheap agricultural products and raw materials, and needing political alliances with the landowners initially to secure markets and investments, both foreign capital and indigenous capitalist groups have, on the other hand, been forced into political opposition to the landowning class by the latter's restrictions on the migration of labour from the countryside, by their reluctance to increase their output by capital-intensive means, and by their refusal to allow any process of land reform that would both permit the emergence of a capitalist class amongst the peasantry and develop a home market for foreign and indigenous industry.

For these reasons, the feudal economic forms introduced by merchants' capital have continued to be reproduced and, despite the adverse economic, political and ideological effects they produce—from the standpoint of either a national capitalist or socialist form of development—the mode of production that sustains these non-capitalist forms is still firmly entrenched, articulating itself with the capitalist mode of production specifically developed by imperialist penetration.

From our examination of the effects of merchants' capital, we can now move on to examine later stages of capitalist penetration, which produced very different effects.[15] We noted above[16] that capitalist forms began to emerge in the non-capitalist formations of the Third World with the qualitatively new forms of penetration they experienced under the dominance of 'competitive capitalism' and 'imperialism' in the nineteenth century. Yet, the problem of how this process occurred has remained unanswered.

How did the elements of a capitalist mode of production develop within these formations? If mercantile penetration within the formations of Latin America, Asia and Africa[17] reinforced the existing unity between direct producers and their means of production, and if the dynamics of the dominant mode produced no tendencies for a long-term separation of direct producers, how were the prerequisites for capitalist production established?

To answer this question, we must turn to the overall effects of penetration under the commodity capitalist stage of the dynamics of the capitalist mode. This form prepared the way for capitalist production but could not—as we shall see—fully establish it.

11 The Effects of Penetration Under the Dominance of Commodity Export

The period of the capitalist formation popularly known as 'competitive capitalism', in which the export of manufactured commodities displaced the previously dominant effects of merchants' capital, has been fully described by many authors, notably by Hobsbawm, in his text, *Industry and Empire*.[1] Hobsbawm shows how, from 1750–1850, the development of manufacturing industry—initially on the basis of cotton merchants' putting-out industries[2]—produced an increasing emphasis on the export of manufactured commodities in those European countries moving towards dominance by a capitalist mode. The period is demarcated by a massive increase in manufacturing exports; these increased in England, for example, by 80 per cent, and colonial trade accounted for one-third of all English commerce by 1770.

Thus, during the eighteenth century, we increasingly find the development of relations between industrialising capitalist and non-capitalist formations being characterised by the export of manufactured goods and the importing of primary raw materials. The English cotton industry is, of course, the prime example: its raw material was imported from sub-tropical or tropical areas, and its products were overwhelmingly sold abroad, particularly on the Indian market; by 1805, it exported about two-thirds of its total output. Indeed, from the empirical evidence available, we can reach the general conclusion that the development of manufacturing industry in Europe during the period up to 1850–70 was, in fact, predicated upon the development of these import-export relations that countries such as Britain had either been able to develop—on the basis of pre-existing mercantile contacts—or to establish through the supremacy of naval power.

If we now try to go beyond this descriptive level, to pose the question of how we can situate the emergence of this dominance of commodity export as an effect of the dynamics of the capitalist mode, of how, during this particular historical period, the tendency of the capitalist mode and the operation of its counter-effects results in the export of commodities—in the same way that, for example, we were able to theorise the basis for the increase in the export of capital during the imperialist stage—we face considerable problems, as we have already indicated.[3] There is, as yet, no rigorous theoretical approach to the analysis of commodity export as an effect of the dynamics of the capitalist mode; and since limitations of space prevent us from investigating this problem further, our analysis will largely remain within the confines of description. We saw above that merchants' capital, through its circuit of $M-C-M^1$, essentially transformed production in the penetrated formation from that of use-value to exchange-value; commodities previously produced either for local consumption or limited trading increasingly became commodities exchanged on an expanding world-market. Through this impetus given to the production of exchange-values, servile relations of production were reinforced.

Penetration during the dominance of commodity export appears to strengthen the tendency of production for excange-value to meet the expanding reproductive needs of the capitalist mode of production, but, rather than reinforcing the non-separation between direct producers and their means of production—as in the earlier mercantile stage—it begins to break this down.

This process is facilitated in modes of production where the determinant relation of production operates through circulation and exchange, or in situations (such as the period of British rule in India) where penetration is accompanied by political control.

Conversely, in modes of production where the exchange of commodities is very limited and is not a means through which the determinant instance operates, or where there is little or no political control by a capitalist power, the process of breaking down the reproduction of the non-capitalist mode of production cannot occur to any considerable extent.

The following exposition will largely develop these points.

THE RESTRUCTURING OF THE NON-CAPITALIST MODE OF PRODUCTION

During the 'competitive capitalist' stage, as we indicated earlier,[4] the tendency of the capitalist mode's dynamics is counteracted, and its enlarged reproduction facilitated, by the importing of raw materials and the development of export markets. Non-capitalist social formations constantly provide the source of the former, and increasingly begin to meet the needs of the latter requirement during this period. Consequently, the circuit of commodity exchange within the non-capitalist mode is increasingly inserted into the circulation of commodities produced by the capitalist mode of production. Raw material commodities required as constant capital in the capitalist mode leave the circuit of commodity exchange in which they were previously confined, and department II commodities previously circulating within the capitalist mode enter into the circuit of the non-capitalist mode, where they are unproductively consumed.

This process, in which the production and consumption of commodities in the non-capitalist mode are increasingly structured by the requirements of capitalist production, does not, however, simply evolve from an earlier form of penetration. Under merchants' capital, the trading of products existed alongside the local system of production, which supplied its own domestic commodities. Under competitive capitalist penetration, these commodities have to be increasingly supplied by the capitalist mode, and more of the raw materials previously used in local production have to be exported to meet the reproductive needs of that mode. Thus, whilst production of raw material goods may increase, it must now do so for the benefit of capitalist reproduction. Concomitantly, the proportion of indigenously-produced commodities relative to imported goods available on the domestic market is reduced. As a result, the reproduction of the non-capitalist mode becomes increasingly geared towards meeting the reproductive requirements of the capitalist mode. This is a qualitatively new form of penetration from that which existed under the dominance of merchants' capital, since it begins to undermine the reproduction of the non-capitalist economic system through trying to inter-relate its circulation of commodities with an external process of circulation determined by the reproductive requirements of capitalist production. This undermining can be more or less successful depending on the presence or absence of particular determinants.

MECHANISMS FOR UNDERMINING THE REPRODUCTION OF THE NON-CAPITALIST MODE

The first and most obvious is that of political control, through the colonial state.This is strikingly exemplified, for instance, in the case of colonialism in India.[5] British policies towards the Indian cotton industry began by securing the protection of European markets from Indian cottons, and this was rapidly followed by acts curtailing cotton production in the colony. Consequently, not only did Britain secure markets for itself in Europe, but, more importantly, by destroying the system of production in India, it created a vast market for exported English cotton commodities. The results for India were disastrous. A mode of production based on a unity between agriculture and artisan industries centred around cotton-production in rural areas was totally undermined—its reproduction was blocked by English economic penetration, which not only restructured consumption to its own reproductive needs, but also began to undermine the existing system of production, to replace it at a later stage (as we shall see) by productive units set up and controlled by British capital. Penetration during the 'competitive' stage can be seen here as laying the foundation for a replacing of one system of production by another, whose reproduction could only ultimately be ensured by the continuing dominance of foreign capital within the economy.

There are, however, less apparent mechanisms by which the reproductive requirements of the capitalist mode of production can come to dominate the reproduction of the non-capitalist mode during the competitive capitalist stage. An inter-penetration of different forms of commodity exchange and circulation can in itself lead to dominance by the capitalist mode's reproductive requirements.

In the capitalist mode, as we know, the determinant instance is the economic, and this operates through an exchange which ties the moments of production, distribution and consumption. This is also the case with other non-capitalist modes—for example, the lineage mode of production;[6] here the determinant relation of production is reproduced via the mechanism of exchange of women, slaves, and élite goods. Here circulation also unifies the various moments within this mode of production. The fact that the initial trading contacts between these two modes of production is beneficial both for the enlarged reproduction of the capitalist mode—since it ensures a supply of elements of constant capital and a provision of markets—and for the reproduction of the determinate relation of production of the lineage mode—since the tribal chiefs' control over distribution and their accumulation of élite goods are both reinforced and extended by this trade—is ultimately a result of their inter-relation being based on an inter-penetration of determinant instances *operating through similar mechanisms*. This enabled capitalist penetration to gain a structural entry in lineage formations. Having achieved this, the capitalist mode then began to undermine the reproduction of the lineage mode. This was achieved by constantly extending its trading spheres to local tribal units of production, thereby objectively reducing the degree to which the unit was dependent for its own reproduction on the supply of women, but, more particularly, of slaves, since in return for material commodities, the European themselves could supply them.

Consequently, through an inter-penetration of modes of production with determinant instances operating through similar mechanisms, the capitalist mode of production has here a greater possibility for ensuring its dominance over the reproduction of the non-capitalist mode.

If, conversely, during the competitive capitalist stage, the reproductive dominance of the capitalist mode cannot be achieved through direct colonial rule or through an inter-penetration of determinant instances, penetration is likely to have little impact on the reproduction of the non-capitalist mode.

This is the case, for example, with formations dominated by the Asiatic mode of production, which were not colonised until the late nineteenth century, and whose determinate relation of production required only a limited circulation of commodities. Production in the agricultural communities remained largely confined to use-values for local consumption, and the commodities that were produced for exchange were collected and distributed via the state to other communities, or externally to neighbouring kingdoms; such exchanges were, however, limited to providing those goods that could not be adequately supplied by the latters' production. For these reasons, capitalist penetration under the dominance of commodity export could not insert itself into the circulation process in the same way that it could, for instance, in the lineage mode, and the capitalist mode of production had to wait until a later stage in its dynamics—that of imperialist penetration—before it could begin to insert itself into the reproduction of the Asiatic mode.

ECONOMIC EFFECTS

Where capitalist penetration can insert itself into the reproduction of the non-capitalist mode via the mechanisms we have outlined above, it produces a series of general economic effects, which can be summarised here.

Production becomes progressively directed towards the needs of enlarged capitalist reproduction, as agricultural and raw material products previously consumed in the non-capitalist formation enter into the circuit of the capitalist mode. Consumption becomes increasingly based on commodities exported from capitalist social formations, and less on indigenous production. Concomitantly, the breaking down of the unity between production and consumption has profound effects on the patterns of distribution of the surplus labour realised in the non-capitalist mode of production. Whether it be a result of changes in the existing system of production, or the restructuring of consumption towards imported goods, there is a tendency both for the amount of surplus labour distributed and the use of productive labour to decline.

This can be illustrated by various examples from British colonialism during this period. In Egypt, British demand for cotton was rapidly accelerating by the 1850s, and this led the Egyptian feudal landowning class to increase the amount of land devoted to cotton cultivation, thereby throwing the peasantry off the lands they previously cultivated, and reducing the surplus labour formerly distributed to them in the form of means of production (tools, seed, and so on). Similarly, by destroying the Indian cotton industry, British penetration produced massive unemployment amongst rural artisans and tenant farmers whose productive labour could no longer be utilised, and whose families and communities could no longer exist on the surplus labour realised from cotton production.

In both these cases, the effect is the same: forms of distribution based on pre-existing systems of production are transformed, and the result is that they are no longer able to meet the consumption requirements in the non-capitalist mode. The pre-existing unity between production and consumption in rural areas begins to be broken down and this results in the development of a migration from rural to urban areas; rural workers begin to be separated from their previous means of production and, entering the town with nothing to sell but their own labour-power, they provided the earliest concentrations of labour for the development of indigenous capitalist production.

Yet, this process was limited during the commodity export state, primarily because the rural economy was, in many cases, able to restructure its production and consumption to meet the burdens resulting from the decline in production; the peasantry also often migrated from one area to another, to regions less influenced by the demands of capitalist reproduction. It was not until the stage of imperialist penetration, when the export of capital required a massive separation of direct producers from their means of production in order to create the bases for indigenous capitalist production, that these processes of separation from the land and means of production and of migration to the cities were fully developed. In this stage, the non-capitalist mode of production had to be destroyed and replaced by a qualitatively new system of indigenous capitalist production, creating the articulation of modes of production that currently determines the structure and reproduction of the social formations of the Third World. It is this process of destruction and replacement that we examine in the next chapter.

12 Imperialism and the Separation of Direct Producers from their Means of Production

In an earlier chapter, we analysed imperialism as a particular stage in the dynamics of the capitalist mode, characterised by a number of structural features depending (ultimately) on the level of subsumption of the productive forces under capitalist relations of production. We showed how this gave rise to the possibility for exporting accumulations of finance capital, and, how, as opposed to earlier forms of capitalist penetration, the dominance of this effect was specific to this stage.[1] In this section we will be concerned with the results of this effect for Third World formations.

THE NECESSITY TO DEVELOP CAPITALIST RELATIONS OF PRODUCTION

We indicated earlier how, during the latter part of the nineteenth century, the possibilities for exporting the finance capital accumulated during the imperialist stage were increasingly limited by blocking on the part of other capitalist formations, and how this capital subsequently entered the colonial areas established under 'competitive capitalism', and the politically independent countries of Latin America. The extent to which capital was exported on this basis depended, of course, on conjunctural features,[2] but its basis ultimately lay in the uneven emergence of the stage of finance capital within the dynamics of different capitalist modes, and the extent to which capital export to other capitalist formations was restricted at the very moment when it emerged as the dominant form within the most advanced capitalist mode.

Forced to move outwards to non-capitalist formations, imperialist penetration was, consequently, faced with modes of production that it had utterly to transform. Capital will only be invested if it produces an increase in the rates of profit for the sector from which it originates—either by enhancing its own enlarged reproduction, or by realising profits from the new units of production it establishes in the penetrated formation. This task could only be achieved if the capitalist formation could succeed in cornering and protecting particular areas of the continents of Africa, Asia and Latin America (either through enforced trading agreements, or political annexation). Such areas could then exclusively provide raw materials for the enlarged reproduction of the capitalist mode, markets for the latter's commodities, and units of production that could yield high long-term rates of profit through the extraction and processing of raw materials for the foreign market, and the production of commodities for a developing consumer and capital goods market in agriculture.

All these developments, however, were predicated upon one factor—*the development of capitalist forms of production in non-capitalist social formations*.

If penetration under the dominance of capital export was to go beyond the limited effects of the commodity export stage and transform production, as was now demanded by accumulations of capital seeking

productive investment, then this required the development of specifically capitalist relations of production. Existing non-capitalist forms of production—structured, as we have seen, by relations of production that produced a dynamics limiting the development of the productive forces—could not provide an adequate basis to meet the increasing agricultural and raw material needs of capitalist social formations; Moreover, the control established during the commodity export stage over the circulation of commodities had proved to be increasingly inadequate to meet these needs. More importantly, the required increase in profit rates—the rationale of capital export—was predicated upon

enlarged reproduction, the preserve of the most advanced form of production for increasing the productivity of labour then known, namely capitalist production. Similarly, any extension of the internal market could only occur if workers were separated from the systems that had previously supplied their subsistence goods, if they received wages for labour-power used in units of production and spent it on goods produced in these units. Consequently, the aims of capital export—both in terms of the realisation of profits and the structuring of production to meet the reproductive needs of more advanced capitalist modes of production—could only be realised through the creation of capitalist relations of production in formations that remained dominated by non-capitalist modes. This had to be the main objective of imperialist penetration.

IMPERIALIST PENETRATION AND THE DOMINANCE OF NON-CAPITALIST MODES OF PRODUCTION

Tracing the genealogy of the elements—capital and labour-power—that combine to form capitalist relations of production, we analysed earlier how capitalist production is impossible without the separation of the direct producer from his means of production; this process ultimately determines the existence or non-existence of a capitalist mode of production. Yet, as we have noted, for many of the non-capitalist modes of production that confronted imperialist penetration—such as the mode we examined—the reproduction of their determinate relation of production required a constant unity between labour and means of labour, and there were no tendencies in the articulation of their dynamics towards any separation. This is obviously not applicable to all non-capitalist social formations. To argue that there were no capitalist tendencies emerging in the dynamics of the semi-feudal modes that dominated Latin American formations during the nineteenth century, or that a formation such as India did not reveal any tendencies towards separation prior to British penetration without a rigorous analysis of their dynamics would be misleading. Our object is not to provide a comprehensive analysis of the dynamics of all known non-capitalist modes—which would obviously be impossible here; the main points we are stressing are: first, that imperialist penetration has a specific effect, in that it necessitates a separation of direct producers from their means of production; and secondly, that this effect has to be analysed in relation to the dynamics of the mode of production it penetrates; only on this basis can we concretely analyse the process by which imperialism set about separating producers from their means of production in particular non-capitalist social formations.

MECHANISMS FOR SEPARATING DIRECT PRODUCERS FROM THEIR MEANS OF PRODUCTION

Having reached this point, we can now turn to the mechanisms whereby imperialist penetration, faced with the necessity of creating capitalist relations of production, attempted to separate direct producers from their means of production in the non-capitalist formation. In the social formations that it penetrated, annexed and protected, imperialism established a colonial state whose primary economic task was to undermine the existing system of production that preserved the unity of the direct producers with their means of production.

Whereas penetration under the dominance of commodity export simply attempted to insert itself into the reproduction of the non-capitalist mode, imperialist penetration began to destroy this mode by attacking the very means by which it was reproduced, *by undermining the reproduction of its determinate relation of production.* In much the same way, for example, that the political alliance between feudal landowners and an emerging capitalist class formed the basis from which state power was able to separate direct producers from their means of production in the transition from feudalism to capitalism,[3] imperialist control of the colonial state was able to block the reproduction of the existing non-capitalist mode by restricting the ability of the dominant class to extract surplus labour, through political means—by depriving this class of its hold over state power, or (and) by destroying the system of production from which it extracted surplus labour.

But how, specifically, did the colonial state achieve these ends? To answer this, we need to examine the major mechanisms through which the colonial state typically brought about a separation of direct producers. The main *economic task* of the colonial state was to create a labour-force in those sectors in which finance capital could be most profitably invested. In the late nineteenth century, these were either the raw material extractive or agricultural cash-crop sectors; it was not until a much later stage that processing or import-substituting industries were set up in colonies and ex-colonies.

Direct producers could be separated from their means of production, from their places within the pre-existing traditional agricultural division of labour, through a variety of strategies, all necessitating intervention by the colonial state apparatus.

Taxation

This could either be in kind or—following the introduction of European coinage—in money, A wide variety of taxes were levied on agricultural producers—on their land, products, transport systems, families, etc.—with the overall object of undermining the producers' ability to exchange their commodities and gain a return adequate for their reproduction, thereby forcing them to sell their labour-power to plantations or mines to ensure their physical survival.

The trading companies that had pre-existed the colonial state in Africa and Asia had, of course, tried to achieve a similar result; increasing output in an attempt to undercut their European competitors, they had tried to rationalise production by introducing a very limited division of labour amongst agricultural producers, but their only effective weapon had been the supply of credit, and this produced extremely limited results in most non-capitalist formations prior to imperialist penetration. With the colonial state, however, the situation was very different, since it could enforce rapacious forms of taxation with the sanction of expropriation. Whether it was applied to rubber producers in Malaya, ground-nut farmers in Senegal, or to rice cultivators in Indochina, taxation had one primary object—to force labour-power out of its traditional pursuits into colonial mines and plantations.

Overall, however, this policy was not very successful. Resistance to tax-collection and the state's inability to organise collection in areas it had only recently annexed were accompanied by agricultural communities increasing their output and distributing the returns from exchange amongst the community, with the result that even the poorest producers were able to survive—more or less.

Land redistribution

A second strategy was more direct. Unable to 'persuade' direct producers to abandon their agricultural communities through the use of taxation, the state could forcibly expropriate the peasantry's lands, handing them over to European companies who would cultivate the crops needed in the home country, and provide the state with a constant source of revenue. Reducing the land available for indigenous cultivation would also, of course, ensure that the peasantry could not grow enough crops to meet taxation needs and survive, since, for many communities, this would now become physically impossible; consequently, their members would be forced to migrate and sell their labour-power.

In many colonies, the changes in land tenure were extensive, and produced disastrous results for the peasantry. In Indochina, for example, French colonialism undermined the agricultural communities by introducing large rice estates which were sub-divided into smallholdings worked by tenants who had to pay 40 percent of their crops as rent to the state. Not only did this process result in a total dependence on the French (or, in some cases Vietnamese) landowner for loans to ensure survial, but it also forced many peasants off the land and into rubber plantations and mines (particularly in N. Vietnam) after the 1914–18 inter-imperialist war. Perhaps the best-known case of land redistribution, however, was that undertaken by the British colonial state in India. In addition to the transformations produced by the destruction of the cotton industry and the failure to maintain canals and irrigation channels in the agricultural sector, British rule also totally destroyed the existing system of land tenure, replacing it, as Marx stated, either by 'a caricature of large-scale English landed estates', or by 'a caricature of small parcelled property'.[4] Beginning with the *zamindari* system, the state expropriated peasant land in favour of the Mohammedan tax-collectors—who then had to give eight-tenths of their rent to the state—and then slowly alienated it into the hands of British financiers during the early nineteenth century; the land was subsequently organised into large estates. Following this, peasant lands were subdivided into small plots and heavily taxed, thereby forcing a dependence on local usurers and colonial loans at rapacious rates of interest. Both these systems (and particularly the former) led to a massive separation of direct producers from their traditional means of production. Whether the peasantry was deprived of its land by the imposition of estates that increasingly demanded rationalised production and wage-labour, or by the introduction of small plots that became more and more unfeasible, the result was the same—labour-power was 'freed' from the agricultural community to work in the plantations or urban industries developed by British finance capital.

These examples could be extended—to British-dominated estate cotton cultivated in Egypt, to French destruction of Algeria's clan and tribal systems of production, and so on. In each case, the colonial state's transformation of land tenure during the imperialist stage, or its preparation for this during the competitive phase, had one major result: the forced creation of a labour-force for an emerging capitalist mode of production.

Infrastructural projects: railway construction

One of the most striking aspects of the early colonial period was the construction of railways, both in Africa and Asia. In the former, for example, whilst in 1860 only 455 kilometres of track had been constructed, by 1910 36,854 kilometres had been built; in Asia, the increase over the same period was even more marked, from 1,393 to 101,916 kilometres.[5] This process often appeared quite irrational, in that railways were built in situations where there was no readily available labour-power, nor any pre-existing capitalist economic infrastructure, nor an adequate justification for the extension of track in relation to the transporting, exporting and importing of goods. There are many such examples that cannot be evaluated in terms of economic rationalism—the Congo railway built by French colonialism, the plethora of railway networks in India, German railway construction in Turkey, and so on. Nor, indeed, should they be evaluated in such terms, since this was not their object; rather, they were explicitly aimed at separating direct producers from their means of production, via the forcible creation of a labour-force, tearing it away physically from the agricultural sector to carry out legally-sanctioned forced labour on state-controlled projects. The construction of railways, apart from providing a secure productive investment for finance capital—since the state generally had ultimate control over the project—and opening up markets and new sources of raw materials, provided the basis for a general system of labour recruitment, in which the state supervised a forced migration of labour into isolated pockets of infrastructural development which depended on the very form of wage-labour that imperialist penetration had to create. Thus, the disintegrating effect of removing labour from agriculture was accompanied by the creation of capitalist relations of production. Whereas the previous two strategies we examined were primarily policies of disintegration, railway construction—as with other infrastructural projects, such as the building of ports and roads—not only achieved this, but also established a 'bridgehead' from which capitalist relations of production could develop. Once armies of wage-workers had been created for this purpose, they could then be employed in capitalist units of production that either developed along the route of the railway (processing factories, mines, plantations), or emerged to supply and service construction (timber works, capitalist farms supplying the subsistence needs of the railway workers, etc.).

Despite a constant loss of labour-power returning to its place in the non-capitalist production, and the instability of seasonal migration from one set of production relations to another, the separation of labour forcibly induced by infrastructural works appears to have been more successful in providing the initial impetus for capitalist production than the strategies of taxation and land redistribution.

Forced labour: mines and plantations

Whilst the intensive application of labour-power to infrastructural development heralded the introduction of force into colonial strategies for developing systems of wage-labour, this policy could be continued and massively extended by a forcible removal of producers from agriculture into the newly-emerging export-sector. Initially, mines, and then, increasingly, plantations received their labour-force through state-controlled forced conscription, particularly in the first decade of the twentieth century.

This proved to be by far the most successful policy. The strategies of taxation and land redistribution enabled the non-capitalist mode to reproduce itself, even if on a much more limited scale, and they provided no firm guarantees that producers displaced from the agricultural division of labour would migrate as wage-labourers to distant capitalist units of production. In fact, in many areas, non-capitalist modes remained impervious to colonial strategies that were not implemented through force. Only when these were applied did a capitalist mode of production begin slowly to emerge, and even then, as we shall see below,[6] its enlarged reproduction was constantly restricted by its articulation with a surviving non-capitalist mode of production.

Thus, of the four main strategies adopted by the colonial state to create the preconditions for capitalist production, those most likely to succeed were attempts made *to forcibly break down* the non-capitalist mode. In every non-capitalist social formation that it penetrated, imperialism, via the mechanism of the colonial state, adopted these strategies in varying combinations and degrees. All colonial policies directed towards the creation of a capitalist labour-force embodied them in different concrete forms.

IMPERIALISM AND TRANSITION

The form of capitalist development that emerges from the breakdown of the non-capitalist mode is distinct. Apart from the limitations placed on its penetration of agriculture by its co-existence with non-capitalist modes or divisions of labour, it is also confined to particular sectors by the reproductive requirements of the industrial capitalist mode of production—as we noted earlier.

The transitional period that imperialist penetration produces is, therefore, characterised by an articulation of modes or divisions of labour in which *the development of the capitalist mode is itself necessarily restricted.* These processes of articulation and restriction are the subject of our final chapter.

The questions we will be posing here are: how can we analyse this particular articulation of modes of production? How does it determine the structure, reproduction and development of the social formation it dominates? How can we analyse the effects on this social formation of imperialist penetration at the economic, political and ideological levels? How does imperialism intervene in the different instances of Third World formations to guarantee the reproduction of the restricted and uneven form of capitalist development that it requires for its own reproduction?

13 The Emergence of an Articulation of Modes of Production and its Effects on the Structure of the Third World Formation

The displacement forcibly induced by imperialist penetration blocking the reproduction of the determinant instance of the non-capitalist mode produces a series of effects that give the transition to capitalism its particular *restricted and uneven form* in Third World formations. In order to analyse these effects, we must initially recall two concepts, *subsumption* and *dislocation.*

Uneven subsumption

As we pointed out in previous chapters, any transition to capitalism is characterised by the establishment of capitalist relations of production in a situation where these relations cannot immediately transform existing techniques of production contained in the division of labour. Initially, therefore, the latter is only *formally subsumed*[1] under capitalist relations. This phenomenon, which characterised the European transition from feudalism to capitalism, is induced in non-capitalist formations by imperialist penetration. Yet, it takes a specific form: the pre-existing divisions of labour are transformed in a very uneven manner, both because imperialism, in the colonial period, focused its attention on those sectors that were vital for its own reproduction, and because, in many formations, it found itself unable totally to break down the non-capitalist mode of production. We thus have an economic structure characterised by an uneven subsumption of non-capitalist divisions of labour under increasingly dominant capitalist relations of production. In some sectors, non-capitalist relations and techniques are retained or slightly transformed; in others they are changed, yet the techniques retained; whilst in the remainder both techniques and relations are replaced by capitalist relations and an increasingly capitalist division of labour. At the most general level, this constitutes the economic base of the transitional period in Third World formations. It gives rise to a particular class structure and specific forms of state and ideology, which—as with the economic structure itself—are governed both by the changing requirements of industrial capitalist production on a world scale, and by the continuing reproduction of elements of the non-capitalist mode of production, which (as we shall see) effectively restricts the development of capitalist relations of production in the agricultural sector. This 'economic base' and its determinants are examined below, but before turning to this, we should briefly note a further effect of imperialist penetration, namely its production of dislocations between the instances of the colonial social formation.

Dislocation

In our earlier analysis of the structure of the social formation,[2] we showed how the predominance of one mode of production sets limits on the extent to which one practice can intervene in another. When a new mode of production emerges to dominate and restrict the reproduction of the previous mode, it becomes increasingly the case that the forms of ideology, state and institutions appropriate to the previously dominant mode no longer correspond to the reproductive requirements of the newly emerging mode of production; economically, the latter requires ideological and political forms that are at variance with the non-capitalist forms that continue to be reproduced relatively autonomously in the transitional Third World formation. Thus, there is not only a dislocation between the dominant mode of production and the ideological and political requirements for its reproduction, but there are also dislocations between and within all the instances of the social formation, resulting from the degree to which imperialism can successfully penetrate these levels. For example, the changes required by imperialism in the political representation of different classes and fractions of classes in the state may be established more effectively and rapidly than the ideologies required to legitimise this change in state power. Or, again, at the ideological level, in particular areas such as the educational system or the administrative apparatus of the state, imperialism may be able to displace existing ideologies more effectively and rapidly than it can others, such as religious or familial ideologies. Here, rather than transforming these forms, colonialism often tried to utilise them to achieve its reproductive requirements.

We will develop these points later. At this stage, we simply want to indicate that the structure of Third World formations can be analysed by focusing on these dislocations between and within instances, *dislocations which can only be examined by a dual reference: to the structure of the pre-existing mode of production, and to the reproductive requirements of the newly emerging capitalist mode of production*, which is itself governed by the changing requirements of the world capitalist economy.

Having outlined the two concepts of subsumption and dislocation, which are crucial for analysing the social formation, we can now go on to examine the basis for this formation—namely the specific nature of the economic structure created by imperialist penetration.

THE ECONOMIC STRUCTURE OF THIRD WORLD FORMATIONS

Restricted and uneven development

Under imperialist penetration, the most thorough transformation of the pre-existing divisions of labour occurred in those sectors that provided the agricultural goods or elements of constant capital required for the process of production in industrial capitalist modes. The separation of direct producers from their means of production here was total. The most advanced capital-intensive techniques were imported from imperialist countries rapidly to augment the rate of surplus value. The degree to which imperialist penetration transformed economic sectors other than those which it required for its own reproduction was, however, extremely variable.

With the exception of the phenomenon of import-substitution, or the production of luxury (department 11a) commodities for the consumption of those classes that acted as supports for imperialist penetration, or the extension of artisan production in commodities that did not compete with imports from imperialist countries, capitalist relations of production did not extend beyond the two major capital-intensive export sectors during the colonial period. Consequently, alongside sectors in which the divisions of labour were rapidly transformed, the spread of capitalist relations was effectively restricted to those areas that would not compete with industrialised capitalist imports. Here the subsumption of the division of labour was much less developed. We can briefly indicate this by analysing these sectors.

Perhaps the least transformed of these labour processes were those that emerged from a union of artisan industry under the dominance of a merchant capitalist emerging during commodity export penetration. This early form of capitalism, in which the local merchant or trader gradually assembled in one unit those to whom he had been selling raw materials and from whom he bought a finished commodity, was a frequent occurrence in Third World formations during the colonial period. Since it simply brought together workers who remained in possession of their means of production, and existed in sectors that foreign capital or commodities did not penetrate, this extension of artisan industry produced little or no transformation in the existing labour processes. Certainly, it extended wage-labour, but the techniques and conditions of production remained essentially unchanged. Furthermore, its position was absolutely determined by the rhythms of imperialist penetration. The process of urbanisation produced by the separation of direct producers in the agricultural sector created an urban market for its products, and it also retained its rural markets and exported commodities. Yet, in all these activities, it could potentially be restricted by imported commodities, which could always enter the market and undercut home-produced commodities.

The position of these merchant-artisan firms in Third World formations is vividly illustrated by their reactions to periods of crisis in the advanced capitalist economies, such as the 1930s depression or the First and Second World Wars, when the degree of imperialist penetration was necessarily lessened. In most cases, these firms were able to supply the commodities previously imported; they experienced considerable increases in profit and capital accumulation; some even began to decrease their level of labour-intensitivity.[3] Yet, with the return of foreign commodities and investment, they were again forced back into those sectors in which they would not compete with more advanced capitalist units of production.

Such then was—and still is—the position of merchant-artisan industry—restricted by imperialist penetration to enclaves producing goods for local consumption, and essentially labour-intensive. The subsumption under capitalist relations of production hardly goes beyond the formal level.

To turn to other sectors: when we come to examine the class structure and alliance and opposition of classes that is formed on the basis of the restricted and uneven economic structure characteristic of Third World formations, we will see that a certain amount of surplus value and surplus labour is not expropriated externally, but distributed to the major class supports of imperialism, such as the comprador-capitalist, financier, and landowning classes. This distribution provides the basis for an internal demand, largely for luxury (department 11a) goods, which is not only met by imports, since it is often cheaper for foreign firms producing these goods to invest in units of production in Third World economies. This is frequently done in co-operation with local capital. Several Latin American and Asian countries, for example, have attempted to extend capitalist relations of production in the industrial sector by maximising the amount of surplus product distributed to the comprador and landowning classes, and encouraging foreign capital to invest in this sector.[4] Ultimately, of course, this strategy founders when the limits of the internal market for department 11a commodities are reached, but, nevertheless, particularly when foreign capital equipment and techniques are imported, these industries can produce a real transformation of existing labour processes, to a much greater degree than is possible in the merchant-artisan sector, yet never attaining the degree of mechanisation reached in the raw material extractive sector. In his text, *Marxism and Underdevelopment*,[5] Geoffrey Kay is quite right to point out that any application of the concept of 'intermediate technology' to this sector totally misses the point, since both foreign and local capital are primarily interested in producing 11a commodities under conditions which enable them to attain the highest rate of profit possible whilst the internal market lasts. The production and distribution of luxury goods is thus another means of transforming the existing division of labour, but its possibilities are severely limited by the characteristics of the internal market.

A further means of achieving the same objective is—as is well known—the strategy of import-substitution, the classic cases of which are several Latin American countries in the 1950s and 1960s.[6] As we indicated earlier, the limitations of such a strategy are clear.[7]

Industries set up tend to produce luxury commodities, and this reinforces the position of the support classes. They have to rely increasingly on imported capital equipment and raw materials, a process which constantly erodes foreign exchange savings, unless the state gains increasing control over the export sector. Finally, import substitution must ultimately, of course, reach the limits of its national market, which, for luxury goods, may not be very large. Thus, the degree to which import substitution promotes an increasing transformation of the division of labour in a capitalist direction is limited. Import substitution generally produces highly capital-intensive units, with complex technology imported from the industrial capitalist economies. These take the form of large-scale units in particular consumer commodity sectors, whose linkages with other sectors are usually minimal. The introduction of a relatively advanced capitalist division of labour is, therefore, generally confined to these sectors. Consequently, import substitution has tended to reinforce rather than lessen the uneven subsumption of the productive forces in the industrial sector of Third World formations.

From the above, we can see that the effect of imperialist penetration on the division of labour is to promote a *restricted* development—restricted in the sense that the extension of capitalist relations of production into sectors occurs only when they are either: (1) crucial for industrial capitalist reproduction, or (2) sectors that do not effectively compete with imported commodities, or, (3) provide a means for strengthening the industrial capitalist state's control over the Third World formation—through import-substitution, through the extension of department II manufacture controlled by foreign-capital, co-owned by foreign and domestic capital, or established by comprador capital largely for export, etc.[8] Despite the conclusions of such writers as Warren, with their contention that dependent industrialisation in specific sectors of the Third World formation heralds the eclipse of imperialism, it is undoubtedly the case that industrialisation will remain restricted in this way. The possibilities of Third World formations embarking on an 'independent industrialisation', on a process of enlarged capitalist reproduction based on an internally-generated or externally non-exploitatively financed investment in department I of production to promote industrialisation geared towards the domestic market, are remote.

We can conclude, therefore, that *in the economic structure of Third World formations, capitalist relations are formally established in a number of sectors, but the existing divisions of labour in these areas are transformed in differing degrees*. The subsumption is uneven because capitalist penetration is necessarily restricted. As we shall see, this uneven and restricted development has important implications for the class structure that results from imperialist penetration.

We cannot develop these points any further, however, without examining a further major determinant of the process of uneven and restricted capitalist development—namely the barriers put up against the extension of capitalist relations of production by the continuing reproduction of elements of the previously dominant non-capitalist mode of production.

The reproduction of the non-capitalist mode and its elements as a barrier to capitalist penetration

In an earlier chapter, we analysed how different forms of penetration either (in the case of merchants' capital) reinforced or introduced non-capitalist modes of production, or (in the case of commodity export) utilised the determinant relations of production of these modes. To establish its dominance over the non-capitalist mode, *capitalist penetration utilised the existing mode only to have to attempt to destroy it at a later (imperialist) stage of penetration.* Yet, because of its past actions, and because of the resistance put up against imperialist penetration by the non-capitalist mode (depending, of course, on the particularity of its dynamics), either this mode or elements of it continue to be reproduced—even when the capitalist mode becomes dominant. In much the same way that, for example, elements of the feudal mode of production continued to be articulated when capitalist modes of production became dominant in W. Europe, and placed restrictions on the utilisation of its surplus product in the form of surplus value, so do elements of the previously dominant mode continue to restrict the development of capitalist relations of production in the transitional formations of the Third World. Thus, *in addition to the restrictions imperialism itself imposes on the form of capitalist development, this process is also reinforced by the continuing existence of elements of the non-capitalist mode.* We can briefly illustrate this point with two examples.

As is well known, capitalism has been remarkably unsuccessful in penetrating the agricultural sectors of Latin American economies. The path to capitalist agriculture through the differentiation of the peasantry and the creation of a home market for national industries has been blocked by the existence of the hacienda system and the political power of the landowning class. The latter, dependent upon a peasantry tied to their agricultural estates[9] have restricted attempts to enact capitalist land reforms which, by redistributing estate land, would lead to the emergence of a capitalist class in agriculture. Such a solution, which could produce a fairly rapid development of capitalist relations of production throughout the economic structure, is restricted by the class alliance that holds state power. The landowning class, with its comprador and financier supports, is continually able to over-ride the demands of the national capitalist class for an extension of capitalist relations (although there have, of course, been periods when this resistance has been temporarily overcome).[10]

This phenomenon results from a past form of penetration, under the dominance of merchants' capital, creating a semi-feudal mode of production, and this then being reinforced during the period of commodity export (in the eighteenth and nineteenth centuries), thereby creating a structural barrier to further development during the imperialist period. Imperialist penetration requires a separation of direct producers from their means of production, and the creation of a differentiated capitalist class and proletariat in both the industrial and agricultural sector; yet the dynamics of the mode of production created and/or reinforced by previous forms of penetration render this impossible, unless the opposition of the alliance of classes on which this mode depends can be overcome by those classes whose actions are limited by this dominant alliance (e.g. the indigenous capitalist class producing for the domestic market), or who are exploited by it (e.g. the industrial and agricultural proletariat, agricultural tenant farmers, etc.). In Latin American formations, this remains unachieved, and we are continually faced with a situation in which two very different modes of production are reproduced concomitantly. Whilst the capitalist mode of production has gained predominance in industry (in the raw material extractive and processing sectors, department IIa production, and in department IIb production that meets the requirements of the industrial capitalist modes), and is attempting to penetrate parts of the agricultural sector, and whilst imperialist penetration has tried both to reduce the political power of the landowning class and to persuade it to accept at least limited capitalist reforms, it still remains the case that the semi-feudal mode of production continues to reproduce itself alongside the increasingly dominant capitalist mode.

Similarly, if we briefly look at two other non-capitalist modes of production, the lineage and Asiatic modes, we see capitalist penetration utilising and reinforcing their dominant relations of production, only to have to attempt to undermine and destroy their reproduction in the period of imperialist penetration. Despite this undermining, however, elements of these non-capitalist modes continue to be reproduced, even when a capitalist mode becomes dominant.

Take, for example, the lineage mode of production. Whilst the tribal village unit's economic functioning was severely curtailed by colonial policies, it nevertheless managed to retain, throughout the colonial period, the very division of labour that existed under the previous relations of production. Agricultural production for the needs of the village unit remained impervious to capitalist penetration, either in the form of an extension of capitalist relations of production or an extension of the market for capitalist commodities. The unity of production and consumption that characterised the division of labour in the tribal-village unit, and which constituted the basis of the lineage mode of production thus continued to be reproduced alongside the dominant capitalist mode (in plantation agriculture and the raw material sector). Whilst it provided the (usually seasonal) labour-power and often the commodities that enabled this labour-power to be reproduced for the capitalist mode of production, this division of labour nevertheless maintained its own reproduction, despite all attempts made during the colonial period to destroy it.

Having reached this point, we can now see that the extension of capitalist production beyond industrial sectors where it can effectively be dominated by imperialist penetration is *restricted both by the strategies for capitalist development imposed by imperialism, and by the continuing reproduction of non-capitalist modes of production and/or divisions of labour which, both during the colonial period and in the present situation, seriously limit the ability of capitalist production to penetrate the agricultural sector*.

Having outlined these two determinants of the co-existence of modes of production and divisions of labour, we can now go beyond this, by outlining ways in which these modes of divisions are inter-related, at the economic level.

Inter-relations of modes of production and their elements

Example I. The types of labour that exist in Third World formations usually take different and often complex forms, quite unlike those found in an industrial capitalist mode of production. The reason for this is that they are *concomitantly related to two different systems of production*. To take a few examples:

Hilferding remarked in *Finanzkapital* that one of the main reasons for capital export was the availability of cheap labour on a large scale, but what he did not point out was one of the major reasons for this—that the value of the labour-power required for capitalist production is partly met in the non-capitalist mode of production or division of labour, and that, consequently, the cost to the capitalist of maintaining labour-power is considerably lessened. This is particularly evidenced in seasonal labour on plantations where the cost to the capitalist company of maintaining the worker during non-productive periods is met by production in the agricultural sector. Much the same holds for labour which migrates from the countryside to the towns during the non-productive periods of village production in other modes. Again, on a more regular basis, the importance to capitalist industry of the value of its labour-power being partly produced elsewhere is evidenced in manufacturing industries that rely on a daily movement from rural to urban areas to maintain high profit levels. Examining the embryonic department II manufacturing industries dominated by foreign capital in politically independent Africa or Asia, in countries such as Nigeria or S. Korea, reveals that this plays a crucial role. Firms in industrial capitalist countries are increasingly discovering that they can produce the same commodity on a cheaper basis in Third World countries[11] (even allowing for transport costs), and one of the major reasons for this is that the subsistence level required to maintain the worker as an instrument for capitalist production can be met by the labour of the family in the village from which he 'commutes' daily. This situation is only possible because this labour is concomitantly related to different forms of production.

Similarly, in the semi-feudal modes of production of Latin America, it is often the case that, for the same worker, wage-labour combines tribute for the hacienda with production for domestic consumption and the market on the family farm. In this case, we again have an *inter-relation between modes of production represented in different forms of labour, through which the value of labour-power is reproduced.*

As a result of this inter-relation, the capitalist mode of production, as it develops, is able to draw in supplies of labour *within limits set by the reproduction of the non-capitalist division of labour.* If this is threatened, then the supply of labour will be reduced. However, up to this point, the inter-relation provides the basis for a supply of cheap labour, highly attractive to imperialism in its colonial phase. This situation has been transformed to a limited extent in many Third World countries, as raw material extraction and the extension of department II production have developed and required a trained, permanent labour-force, but it still persists in many sectors of production—particularly when the low level of wages which founded the initial basis for investment is perpetuated by state policies and planning that are predicated upon continuing and expanding foreign investment.[12]

This inter-relation of different forms of production manifesting itself in a combination of different types of labour, has important results both for the class structure and for the ideological forms in which the relationship of workers to production is lived. It produces a differentiated industrial proletariat and semi-proletariat, whose conceptions of their place in the system of production and the conflicts endemic to it are qualitatively different from those emerging in previous transitions to capitalism. This is an important point, to which we will return in our analysis of the class structure later in this chapter.

Example II. Artisan industry, as we indicated above, is both extended and restricted by capitalist penetration of the non-capitalist social formation. At each crucial point in its development—the introduction of commodity production, the creation of an enlarged internal market through the separation of direct producers, and (eventually) the extension of its market through the development of capitalist production in agriculture—it can be restricted by commodities produced under more advanced capitalist conditions. Nevertheless, during each of these phases it exists as the result of inter-relations between two modes of production, elements of which continually exist within it. Whilst its production is increasingly geared towards the sale of commodities, its division of labour remains, in most cases, an extension of the techniques of artisan production, with the previously isolated workers increasingly located in one unit of production, under the direction of the merchant who previously bought and sold to them. These industries thus combine a capitalist organisation of their units, induced by foreign penetration and the emergence of a commodity market, with the retention of divisions of labour characteristic of the non-capitalist mode.

Merchant-capitalist artisan production can, of course, be extended into the export sector, and into those sectors that do not compete with foreign capital. In case of the former, as we noted earlier, this extension has usually been the result of a period of isolation (through wars, inter-imperialist crises, changes in the type of commodity extracted);[13] in the latter, it has been confined to sectors processing raw materials or to particular sectors of department II production. In these cases, the division of labour inherited from the non-capitalist mode is transformed, and, rather than combining elements of the two modes of production, these units of production increasingly develop the combinations characteristic of capitalist manufacturing, and (particularly in the case of the import-substituting sector) of machine industry.

Given these limited possibilities for transformation, the capitalist extension of the artisan form, combining elements of two modes of production, necessarily remains a type characteristic of Third World formations. As such it produces a crucial political effect; the restrictions placed on its development by imperialist penetration lead the owners and controllers of the means of production in this sector into political opposition to those classes whose material existence is guaranteed by penetration. This class conflict, which has characterised many Third

World formations both during and after the colonial period (and for Latin American formations since the late nineteenth century) will be examined in greater detail below.

Example III. The co-existence of modes of production in Third World formations greatly enhances *the role of the merchant or trader* which, as we have seen earlier, may either have accompanied the development of commodity exchange in the non-capitalist mode, or been introduced by penetration under the dominance of merchants' capital and commodity export.

The expansion of the home market, the export of agricultural commodities and goods from the national capitalist sector, and the distribution of imported commodities, all provide a basis for a rapid increase in the circulation of commodities, and for an economic strengthening of the role of the merchant class. The latter essentially acts as a 'linkman'[14] between modes of production or divisions of labour, purchasing, for example, a commodity produced in the non-capitalist division of labour (e.g. the tribal village unit, the communal village) and selling it to another trader or to the home market created by the introduction of capitalist production in the urban areas.

This expansion of the trading function provides the basis both for ensuring that non-capitalist production is increasingly directed towards the needs of capitalist reproduction, and for the growth of *a trading class dependent upon the maintenance of the restricted and uneven development that imperialist penetration has produced.* Once again we have, therefore, a superstructural phenomenon dependent upon the transitional economic forms that imperialist penetration has produced.

From these examples, which indicate a number of ways in which non-capitalist and capitalist economic forms are inter-related in the transitional formations of the Third World—in the areas of provision of labour-power, production, circulation and distribution—it seems that they result both from the restricted and uneven development required by imperialism and from the resistance of non-capitalist structures to capitalist penetration; furthermore, the capitalist mode attempts to direct non-capitalist production towards its own reproductive needs by utilising the very phenomena produced by the inter-relation of different forms of production.

An articulation of modes of production

When we outlined the structure of the social formation in an earlier chapter, we noted that the interventions of one practice within another are determined by the limits placed on the extent to which one practice can modify another, these limits being ultimately set by the requirements for the reproduction of the dominant relations of production.[15] Thus, in the capitalist mode, for example, state intervention in economic practice is governed by the requirements of that practice in specific economic conjunctures which exist as particular variants of the operation of the counteracting tendencies to the falling rate of profit. Intervention is forced to operate within these limits, if it is to ensure the continuing enlarged reproduction of the capitalist mode.

In analysing Third World formations, we are concerned with *the articulation of a practice of one social formation within a practice of another social formation*. Each practice attempts to transform the other to meet the reproductive requirements of its own mode of production. Yet its articulation is restricted by the limits within which the penetrated practice must operate to reproduce the mode of production in which it exists. When, however, the reproduction of the dominant mode of production becomes blocked by the emergence of a new mode of production (as an effect of imperialist penetration), the articulation of practices has as its object to transform those appropriate to the previous mode of production, to restructure them in relation to the reproductive requirements of the new mode of production. Thus, for example, the intervention of political practice (through the mechanism of the colonial state apparatus) in the economic instance of the transitional Third World formation aims to transform it (through the mechanisms we have already outlined)[16] to meet the requirements of imperialist penetration in its different stages. On the other hand, this intervention is, as we have argued above, itself restricted by the continuing reproduction of the previously dominant mode of production or of elements of this mode of production.

The conclusion from this is clear. In the case of the non-capitalist social formations of the Third World penetrated by capitalism where, as an effect of imperialist penetration, the capitalist mode is becoming dominant, the articulation of one practice within another is governed both by the reproductive requirements of the capitalist mode and by the restrictions placed on this articulation either by the limits within which the penetrated instance can operate, as set by the non-capitalist mode of production, or by the continuing reproduction of elements of the non-capitalist mode. This point is crucial for our analysis.

In the section above, we have referred to a 'co-existence' of forms of production, 'inter-related' in various ways. It was essential to do this, in order to indicate that the Third World formation contains a combination of modes of production, resulting from imperialist penetration, and that this promotes a process of restricted and uneven economic development. Yet, these notions are essentially inadequate since, whilst they describe what exists (a 'co-existence', an 'inter-relation'), they give us no indication of its determinants. The concept of *articulation* goes beyond this, by relating the transitional combination of modes of production to its determinants: to the reproductive requirements of imperialist penetration and the restricted capitalist development it sets up, and to the resistance of the non-capitalist mode of production and its elements.

What this means, and what we will briefly examine, is that any *economic phenomenon* one analyses in a transitional Third World formation is a phenomenal form, the *determinants of which*—the articulation of a mode of production or its elements in another mode—*are not manifest in its appearance*. Consequently, rather than stating that a phenomenon, such as the different forms taken by labour-service, is the result of the existence of two modes of production and the ways in which they are inter-related, we can now go on to state more rigorously that it is *the result of an articulation of different modes of production, and that this articulation is structured by the reproductive requirements of the capitalist mode of production on one hand and the resistance of the non-capitalist mode or its elements on the other, with both the requirements and level of resistance changing over time.*

Economic phenomena as effects of this articulation of modes of production

Having made this statement, I now intend to examine two specific economic phenomena—indigenous accumulations of capital and urban unemployment in Third World formations—showing how they are the result of an articulation of instances that changes over time, as determined by variations in their two determinants, outlined above. We will briefly conclude this section on articulation by indicating, on a more general level, how two phenomena we descriptively examined above—the forms of labour-service and division of labour that exist in transitional formations—can also be analysed from within this framework.

Example I. Accumulations of monetary capital. The accumulations of capital that combine with newly created labour-power in the sectors which capitalist production dominates are not solely of industrial capitalist origin. Apart from those classes who are involved in the system of imperialist penetration, and can have access to monetary wealth as a result of this, there is a further important source of indigenous capital accumulation operative even in the initial stages of capitalist development. This has its basis in the political and economic dominance of the non-capitalist classes in the previous mode of production being reproduced 'in a new form' in the capitalist mode as a result of these classes having access to the value produced in the capitalist mode of production. In much the same way that the determinant relations of production of the European feudal mode can continue to be articulated within the capitalist mode as a relation of distribution (the landowning class having a legally-defined right to part of the surplus value produced through the extraction of ground-rent), so can the determinate relations of production of non-capitalist modes in Third World formations be perpetuated as a relations of distribution.

Take, for example, the lineage mode of production, as it existed in West Central Africa prior to colonialism.[17] Here the elder extracted surplus labour from the village units in the form of élite goods, in return for which he supplied women and slaves. With the penetration of competitive capitalism, and the threat to the reproduction of the determinate relation of production posed by more direct commodity exchange with the village units, the elder attempted to strengthen his hold over the village units by demanding a larger dowry, representing a greater amount of the surplus product. This reinforcing of the determinate relation of production is the result of an inter-penetration of the economic instances of two modes of production, a process which is limited, at this stage, by the reproductive requirements of the lineage mode. Once, however, the reproduction of this mode is restricted by the enforced decline of village production promoted by the colonial state,[18] the elders cannot extract surplus labour from the villages to the extent previously possible. Yet, because the kinship structure required for the lineage mode continues to be reproduced when this mode is no longer dominant, the young male in the village units must still give a dowry to obtain a wife. He resolves his dilemma by presenting as a dowry saved earnings from the wages he earns as a migrant or seasonal worker in the capitalist mode of production. The dowry, once presented in the form of élite goods, now assumes a monetary form.[19]

Thus, due to the continuing reproduction of kinship relations, with their material base in a reduced lineage mode confined to non-plantation, rural areas, the relation between elders and village units assumes the form, *within the capitalist mode*, of a right of the elder to a part of the value received for the sale of labour-power. We have, therefore, an articulation of a previously dominant relation of production within a mode of production in which it can only operate as a relation of distribution. This articulation enables the elder to accumulate capital which he can then invest in the capitalist mode, combining it with the labour that has been forcibly released from the lineage mode. Thus, despite the restrictions placed on the relations of production that establish their dominance, the class of elders is able to retain its political power through continuing to accumulate wealth in a new (monetary) form, wealth which can only exist as a result of an articulation of the economic instances of two modes of production—an articulation that is made possible by the continuing resistance of the lineage mode to capitalist penetration.

Again, in Latin American formations, despite the resistance put up to capitalist penetration of agriculture by the landowning class dominant in the semi-feudal mode of production, we have a similar process of articulation in the small sectors of capitalist agriculture. Here, the landowners are able to claim a part of the surplus value realised because the legal form of private property developed under the dominance of capitalist production (individual ownership as a legally sanctioned relation) allows ground-rent—the determinate relation of the semi-feudal mode—to be extracted from the producer in an absolute as well as a differential form.[20] Consequently, even given the decline of the semi-feudal mode that supports the landowning class, it will still be able to retain a basis for capital accumulation and economic control in the process of capitalist production through its ability to claim part of the surplus value realised in capitalist production, a claim that is an effect of a continuing articulation of a previously dominant relation of production in a new mode.

From this brief examination, we can draw an important conclusion: that the continuing economic power and political position that previously dominant non-capitalist classes have within the state apparatus in Third World formations is not simply a result of their being involved in the operations of economic sectors promoted by imperialism, but is also a product of the ideologically established 'right' they have to claim part of the value produced in these capitalist sectors. This right, as we have tried to show, can only be adequately analysed as ultimately dependent on the articulation of economic instances characteristic of the transitional period, an articulation that is governed and restricted by the reproductive requirements of the two modes of production that co-exist in the Third World formation.

Example II. Urban unemployment. To provide a further example of economic phenomena being structured by an articulation that is determined by the reproductive requirements of two modes or divisions of labour, we can briefly examine an important characteristic of Third World economies, namely their massive levels of urban unemployment.

We have already outlined how imperialist penetration, through such means as the creation of plantations, the destruction of rural industry, the introduction of cash-crop cultivation for export on semi-feudal estates, and enforced migration of labour, produced a separation of direct producers from their means of production in the agricultural sector. These processes created the essential prerequisite for capitalist production. Yet—with the exception of what we have termed merchant-artisan capitalist production—the sectors in which capitalism developed, both in industry and agriculture, tended to be capital-intensive—a necessary accompaniment of the rapid introduction of the divisions of labour required by foreign investment and its class supports.

Thus, at an early stage of imperialist penetration, an extensive separation of direct producers was enforced, as a prerequisite for enlarged capitalist reproduction. Yet, at a later stage, given the restricted form of capitalist development that ensued in both industry and agriculture, this labour-power could not be fully utilised. Unable to return to the non-capitalist agricultural production which had sustained it, due to the destruction or drastic reduction of this sector, and unable to find employment in industry, this mass of labour-power was forced either to remain unemployed, or to seek employment in the inflated non-productive service sector that has arisen as a result of the above process, in the cities of Third World countries.[21]

We can see, therefore, that this phenomenon can also be analysed as a result of an articulation of instances of different modes of production changing over time.

Furthermore, it is, again, vital for our analysis of the class structure of Third World formations. The confining of labour-power to sectors where it cannot be productively utilised has produced, alongside a differentiated industrial proletariat, what we can term a 'semi-proletariat', sporadically employed in the state sector, the service sector, or in sectors dependent upon foreign consumption,[22] and subject to the most appalling conditions of existence.[23] Politically, the semi-proletariat is an extremely contradictory stratum, having vacillated historically between ideologies as widely disparate as right-wing populism, Bonapartism, and socialism. In many countries, it constitutes a considerable number of the urban population.[24] We will be examining the place of this semi-proletariat in the class structure of Third World formations later in the chapter; for the moment, we should stress our conclusion that the investigation of this class must be approached from an analysis of the determinants of urban unemployment in the transitional Third World formation, and that these determinants themselves are the result of an articulation of instances specific to Third World formations.

Example III. Combinations of labour. As we indicated above, the reproduction of labour-power in Third World formations can be concomitantly related to productive processes in different modes of production. This phenomenon, and the variety of combinations of labour and labour-service that one finds in Third World formations, is, again, an effect of the articulation outlined above. The forms taken by labour are determined by processes which are not deducible from particular combinations as they appear in different sectors. Having made this point, and outlined specific examples of combinations of labour above, We shall here simply indicate the major aspects.

The restricted and uneven development produced by imperialist penetration and by the continuing reproduction of at least elements of the non-capitalist mode, produces forms of labour that unite systems of production whose basic features are quite opposite. These forms are transitory, in the sense that the reproductive requirements of the different systems that determine them change along with different phases of imperialist penetration.

For example, the confining of penetration to the plantation, infrastructure and extractive sectors maintains the workers' links with the non-capitalist mode to a greater degree than a phase of manufacturing industry producing for the home-market in the cities. Again, the development of capitalism in the agricultural sector, with its concomitant processes of differentiation and proletarisation of the peasantry, will produce a further decline in the forms of labour characteristic of the non-capitalist division of labour. Consequently, when we analyse the various combinations of labour that exist in a particular Third World formation, and the effects that these have upon the class structure, ideology, and so on, we must always situate them in relation to the determinants of the articulation that governs them, to the level of penetration that imperialism has been able to attain.

These combinations can take a variety of forms: seasonal capitalist labour whose value is partly reproduced in the non-capitalist mode, combinations of merchant-capitalist artisan labour and domestic agricultural labour, wage-labour on plantations and domestic labour, peasant family capitalist agricultural labour combined with labour-service on latifundia, combinations of labour-service and wage-labour, communal village labour combined with wage-labour, and so on.

In any particular Third World formation, these forms, or even several of them, are *'unified' in the labour performed by an individual worker*. This combination of labour-types produces highly differentiated strata of workers in both industry and agriculture—a feature specific to Third World formations, whose superstructural importance will be assessed later in this chapter.

Example IV. The division of labour. As we pointed out earlier, the subsumption of the productive forces under capitalist relations of production proceeds very unevenly in the industrial sectors of Third World formations. The articulation of the political and economic instances of the capitalist within the non-capitalist mode produces, for reasons we have already outlined, a tendency for the division of labour to remain at the level of simple co-operation in sectors producing for the domestic market. Consequently, we again have here an economic phenomenon in which the elements of different modes of production are unified—a combination of specifically capitalist relations of production with divisions of labour characteristic of artisan production in the non-capitalist mode of production.

This combination is, of course, subject to transformations in the process of articulation of the two modes.

We can see this quite clearly, for example, in the recent movements of foreign capital into the domestic industrial sectors of Third World formations, where, often through co-investment with domestic capital, it is producing a transformation of the division of labour towards that required by machine industry. The result is that domestic industry which attempts to retain its independence from foreign capital is increasingly forced into the least profitable industrial sectors. The national capitalist class increasingly faces the choice either of integrating itself with or becoming dependent on imperialist penetration (through the supply of capital equipment), or of remaining in the backward domestic sector producing for the domestic market.

As a result of this transformation in imperialist penetration of the industrial sector, and the subsequent growth of department II manufacture, development is becoming less uneven within the industrial sector, but, conversely, it is becoming more restricted in the sense that the development of this sector is confined to areas which the industrial states can effectively control, or which do not compete with the latters' marketing requirements, as we noted earlier in this chapter.[25]

Thus, as with the other economic phenomena we have examined, the division of labour existing in the industrial sector of Third World formations is one in which various forms of labour process, from a simple capitalist reorganisation of artisan industry to manufacture and machine industry, co-exist. Within this co-existence, the variations in the extent to which one process prevails over the others is determined by an articulation of instances which changes as its own determinants are transformed through the deepening of imperialist penetration.

The specificity of the economic structure of third world formations

We have argued that there is a form of economic development specific to Third World formations in which the development of the productive forces is restricted to particular sectors of capitalist production. We have analysed the reasons for this restriction as twofold. They relate, firstly, to the reproductive requirements of the industrial capitalist modes of production, and, secondly, to the barriers put up against imperialist penetration by the continuing reproduction of non-capitalist modes of production and/or divisions of labour.

Both of these determinants have changed during the course of imperialist penetration. From initially penetrating the primary, extractive and infrastructural sectors, imperialism has promoted the extension of capitalist relations of production primarily to department IIa, but more recently to some sectors of IIb, producing both for the domestic market and for export. Yet, it has done so in such a way that its control over the industrial sector has been strengthened. Similarly, the restrictions placed on imperialist penetration, most notably in the agricultural sector, by the reproduction of non-capitalist modes or their elements have also changed. Such factors as the increase in production for the market via the expansion of trading links, the capitalisation of artisan industry, the limited expansion of employment produced by the extension of capitalist relations to department II sectors, and the tentative growth of capitalist agriculture in some Third World formations,[26] have meant that the dependence of labour-power on production in the non-capitalist division of labour has, to a limited extent—that is relative to each particular social formation—been reduced.

Despite these changes, however, it remains the case that, because of the restrictions placed on the extent to which the division of labour can be increasingly transformed throughout the economy in the direction of organic compositions characteristic of the industrial capitalist modes of production, the level of subsumption of the productive forces still develops very unevenly. The effect of the continuing articulation of the economic instances of different modes of production and divisions of labour is to produce a specific combination of capitalist and non-capitalist relations of production and productive forces, which blocks the development of the productive forces, by conserving the existing division of labour in some areas, and utterly transforming it in others.

Thus, economic phenomena characteristic of Third World formations, such as the availability of cheap labour, low productivity, high unemployment, gross inequalities of income, subsistence living standards, etc., are specific effects of this articulation, *necessarily combining elements from different modes of production in forms whose determinants are not immediately given within these phenomena as they present themselves for analysis*. The basis for the continuing reproduction of these phenomena must thus be sought within an articulation of economic instances that is both a result of and a condition for the continuing reproduction of industrial capitalist modes' dominance within the world capitalist economy.

Having outlined the transitional economic structure of Third World formations, we shall now examine the characteristics of the superstructure that is determined in the last instance by this economic articulation. In particular, we will focus on the class structure, the political representation of classes and class fractions within the state, and attempt to develop a framework in which the emergence of new ideological forms and the transformation of existing ideologies can be analysed.

THE CLASS STRUCTURE OF THIRD WORLD FORMATIONS

The dislocation of the political and economic instances

Through its enforced separation of direct producers from their means of production, imperialist penetration—via the mechanism of the colonial state—blocks the reproduction of the determinant relation of production of the non-capitalist mode. Consequently, the process of extraction of surplus labour that formed the material basis for the political dominance of a particular class in the non-capitalist mode is undermined by the increasing dominance of capitalist relations of production.

In areas such as the semi-feudal formations of Latin America, where imperialist penetration is not established through a colonial state apparatus, it is impossible for it to block the reproduction of the non-capitalist mode in as direct a way as it can through colonial rule. Consequently, there is a more gradual process of restriction, in which the limited separation of direct producers permitted by the landowning class results from processes that gestate during competitive capitalist penetration—such as the decline of handicraft industry, the development of cash-crop production for export on estates—processes which initiate a labour force whose supply is rapidly increased by the development of the extractive and infrastructural sectors during the imperialist stage. Consequently, the control by the dominant landowning class over the state apparatus is maintained to a far greater degree than under colonial control, both because capital export and investment initially require the support of this class, and because the emerging capitalist mode of production does not really encroach upon the semi-feudal mode until the limits of the domestic industrial sector are reached, and a capitalist penetration of the agricultural sector is required. We thus have a gradual restriction or confinement, rather than a more direct undermining of the determinant relation of production.

Yet, no matter what the relative degree of blocking or restricting of the non-capitalist mode, there emerges a clear dislocation between the political and economic instances of the transitional Third World formation during the imperialist stage. *Whilst capitalist relations of production become increasingly dominant, the political precondition for the reproduction of the capitalist mode of production, namely the holding of state power by the political representatives of the economically dominant capitalist class, can only be established through the use of the oppressive apparatus of the colonial state.*[27] There is, furthermore, no basis for an ideological support for this state amongst the population; rather, the ideologies within which the political authority of the previously dominant classes is represented continue to be reproduced. This is particularly the case in transitional formations without a colonial state, since here the enlarged reproduction of capitalist relations is constantly limited by the continuing political control that the dominant class of the semi-feudal mode is able to exercise within the state apparatus.

Unless imperialism can establish the political dominance of a class or alliance of classes which can gain ideological support amongst sectors of the population and intervene, via the state, in the combination of modes of production to promote the dominance of the capitalist mode, the reproduction of the capitalist social relations necessary for the enlarged reproduction of the capitalist mode cannot be guaranteed. The forms of physical oppression outlined in the previous chapter can establish the preconditions for the development of capitalist production, but its continuing reproduction cannot be guaranteed on such a hazardous basis. It requires both an ideological and a political foundation, a commitment to its adequacy as a superior form of production in the ideologies that structure daily life, and a permanent access to political power to guarantee its perpetuation.

Dislocation and imperialist strategies

Faced with the effects of the dislocation it produced, the strategy of imperialist penetration in its initial phase was twofold. First, it assisted the incorporation of the dominant class or classes into the system of capitalist production by allowing them to retain access to the surplus value realised in capitalist production. By permitting a continuing articulation of non-capitalist relations of production as relations of distribution, imperialism provided a material basis for their continuing political dominance.

Secondly, imperialism attempted to promote politically those classes whose economic dominance resulted specifically from the mode of production that it forcibly introduced, as a result of the qualitatively new class structure that emerged from this mode of production.

These two strategies were not mutually exclusive, and policies pursued during the early imperialist stage generally combined both. To analyse how this occurred, and how imperialism was able to promote the dominance of certain classes over others in particular Third World formations, we must now examine the characteristics of the class structure founded by the economic articulation outlined above.

In what follows, we will outline the major classes and class fractions contained in this structure, indicating how they are founded by the economic articulation. Our object, again, is to produce a general framework in which the class structure of particular Third World formations can be theorised.

The differentiation of the proletariat

Within Third World formations, the co-existence of different forms of division of labour required by restricted and uneven development produces a highly differentiated proletariat. Furthermore, this differentiation varies from one phase of imperialist penetration to another. (i) During the colonial period, capitalist penetration—largely restricted to the extractive and agricultural export sectors—required a division within the proletariat between a small, trained labour force permanently tied to the unit production—to fulfil the 'specialised' technological tasks required by the existence of machine industry in parts of these sectors, notably the raw material extractive sector—and a more general supply of migrant labour, whose value was reproduced in another system of production. These supplies of cheap labour, produced by the action of the colonial state, were thus combined with a small fraction of the newly emerging proletariat, whose wages and working conditions, as required by the technical division of labour introduced by imperialism, were relatively superior to those of migrant labour.

Despite the economic determinist arguments of those who try to analyse this fraction of the colonial proletariat as a 'labour aristocracy', attributing an ideology to it simply on the basis of its more 'advantageous' economic position in the division of labour in the colonial export sector,[28] it seems that the major problem here is rather one of analysing the effect that occupying different places in the division of labour in these sectors has upon the pre-existing overdetermination of ideologies dominant in the non-capitalist social formation. This is a complex problem, which—for lack of space—cannot be rigorously posed here, but, briefly, we should note that the ideological effects of full participation as a wage-worker in a capitalist unit of production, as compared with temporary involvement as a migrant worker are considerable, and have crucial political implications for the analysis of working-class movements during the colonial period.

For example, take the case of one formation—Indonesia during the period 1880–1930. Here, despite the importance of Dutch metropolitan labour in the development of the trade-union movement, an important role in developing the ideology of economic struggle in capitalist units of production was played by the small, permanent, indigenous labour force in the extractive sector. Whilst the struggles of seasonal and migrant labour were expressed in varying combinations of religious ideologies (syncretic forms of animism and Islam) produced in the village units of the Asiatic mode, and realised in sporadic isolated acts of militancy,[29] the trained labour-force in this sector tended to adopt forms of economic struggle more characteristic of those prevalent in units of production in developed capitalist formations. The amalgamation of these two forms is clearly evidenced in the political development of the nationalist and communist movements prior to the independence period.[30] Consequently, as a result of imperialist penetration, we have, on the one hand, a particular place occupied in the division of labour producing a transformation of non-capitalist ideological concepts (e.g. work, power, authority), and, on the other hand, another place resulting in a reproduction of these conceptions in a slightly transformed form. Whilst the case of Indonesia is not, overall, as we have seen, a typical form of imperialist penetration, it nevertheless illustrates a basic differentiation of the proletariat produced by imperialism in the export sector during the colonial phase.

(ii) Alongside this twofold differentiation, there is, as we have noted, a much more gradual proletarianisation of artisan labour in those sectors (domestic merchant-artisan capital) not dominated or controlled by imperialism. Because of the more formal subsumption of the productive forces here, the unity between the worker and his means of production is retained to a much greater extent than in the capitalist export sector. Consequently, the ideological effects of this system of simple co-operation can, again, be clearly demarcated from the two types of labour involved in the more complex division in the export sector. Producing for the local market, existing alongside village agricultural systems, retaining its craft-skills as a precondition for production, and constantly being threatened with unemployment by undercutting from imported commodities, the fraction of the colonial proletariat in the merchant-artisan capital sector tends to combine a rejection of imperialism with a rejection of all forms of capitalist production in general in favour of a simple—yet objectively impossible—return to pre-capital individual artisan labour. This conception, as we shall see below, is very clearly reflected in its forms of political representation.

(iii) We noted earlier that the articulation of modes of production and divisions of labour characteristic of Third World formations, produces, of necessity, a mass of semi-proletarianised workers, centred in urban areas. The ideologies dominating this section of the proletariat are, again, very different from those prevalent in other sectors. Despite the conclusions of several authors, that, because it is 'marginal' to production, this semi-proletariat is inherently 'revolutionary',[31] or that it fulfils a role in the Third World formation equivalent to that prescribed by Marx for the 'lumpenproletariat' in industrial capitalist formations,[32] it is clear that this section of the proletariat is dominated by elements from very different ideologies that are often mutually exclusive and contradictory.

The fact that this fraction has, during the post-war period, played a crucial part in very different political movements, which can be variously characterised as Bonapartist[33] (e.g. Nasserism, Sukarnoism), right-wing populist (e.g. Cambodia under Sihanouk in the mid-sixties) and socialist (the Chinese Revolutionary movement), is primarily a political effect of its necessarily sporadic employment in very different (generally non-productive) urban sectors, the state apparatus, the distributive sectors, trading, tourism and temporary self-employment in the retailing sector. The political ideologies dominating the semi-proletariat cannot, of course, be simply deduced from their differing conditions of work, since they must also contain elements from an overdetermination of the practico-social ideologies dominant in the non-capitalist formation, the ideologies dominant in the state apparatus, the ideologies prevalent in the agricultural sectors in which semi-proletarianised labour may seasonally return, and so on. Nevertheless the reactions of this fraction to these ideologies are necessarily influenced by their conceptions of work, of the nature of struggles in the unit of production, and of their relations to the section of the class dominant within their sector of production, and this provides us with what seems to be the major distinguishing feature, ideologically, of this fraction of the proletariat. Whilst there is a degree of common experience of struggle within similar units of production amongst other sections of the proletariat, this is clearly not the case with the semi-proletariat; at one historical moment during a phase of state expansion, sections of it may be forced into employment in the service sector, to be dominated by conceptions prevalent amongst the collective labour in the export sectors; whilst, at another moment, inflated by a rapid increase in rural migration, the majority may be forced into the parasitic tourist sector, to work as peddlers, pedicab drivers, beggars, etc. This lack of permanent employment in one particular sector has a profound political effect. In one period, sections of the semi-proletariat can, for example, be attracted to socialist-based movements emerging from the urban and rural proletariat; in another period, they could equally well rally behind a populist movement based on indigenous industrial capital and rural smallholders; in other periods, they can even aspire to the political ideologies of the urban petit-bourgeoisie. The contradictoriness and ambivalence of political movement amongst this large and powerful section of the proletariat[34] must, therefore, be situated primarily within the place it necessarily occupies 'on the periphery' of production. Perhaps more so than other exploited classes in the Third World formation, its ambivalent place in the economic structure seems to be the key determinant in its political ideological formation. Consequently, its political direction is profoundly conjunctural.

(iv) This four-fold differentiation of the colonial proletariat is further compounded by the extension of capitalist production to departments 11a, and (gradually) to 11b, that occurs during the period of political independence.[35]

In both these departments, the technical division of labour generally requires—to a far greater extent than during the colonial period—a mass of labour trained in the very basic skills needed to operate the technology introduced into these sectors. This is less so with the extension of the infrastructure that has occurred with the development of department II production, and for manual employment in the state sector, but for the import-substituting, luxury goods, and domestic commodity sectors, this generalisation appears to be valid. Furthermore, co-existing alongside this emerging semi-skilled fraction, there has been a tendency, particularly in the more capital-intensive importing-substituting sector, for there to develop a small, highly trained group of workers, with the specific task of ensuring the maintenance of imported technology. Consequently, at the most general level, we have a situation in which the proletariat is now further differentiated, with the mass of workers in the newer industries occupying the most basic semi-skilled positions in a capitalist division of labour.

This compounding of differentiations upon existing colonial demarcations produces a structured proletariat that is *specific to Third World formations.*

The uneven and restricted forms of capitalist development produced by the articulation of modes of production and their elements during the colonial stage of imperialist penetration produces a proletariat differentiated into trained permanent, unskilled migrant, artisan-capitalist and semi-proletarian fractions. The extension of capitalist production that occurs during the post-independence period, during a qualitatively different phase of imperialist penetration in which the overall subsumption of the productive forces becomes less uneven, compounds this set of demarcations by adding to them an extension of the semi-skilled fraction of the proletariat. Consequently, co-existing with those ideologies that emerged in industrial sectors during the colonial period, we now have an extension of the conceptions of work and of economic struggle generated in specifically capitalist units of production, to which sections of the proletariat are increasingly tied on a more permanent basis. Within the overdetermination of economic ideologies existing within the proletariat, there appears, therefore, to be an increasing dominance—within specific sections of this class—of conceptions of economic struggle appropriate to more technically advanced capitalist units of production over those generated or perpetuated during the colonial period. As we shall see below, this has considerable importance for the process of political representation in the post-independence period; for the moment however, we should note that it produces a potential basis for the emergence of a 'labour aristocracy', combining the relatively more economically advantaged sections in the imperialist dominated extractive processing and consumer goods sectors. These constitute a 'labour aristocracy', not in any simple sense of their economic ideologies being deducible from their relatively advantaged wages and conditions of work, devolving from 'super-profits', but rather that the notion of their place in the division of labour as relatively more skilled functionaries, forms a crucial aspect of their economic ideology—crucial in the sense that this aspect has been important in ultimately determining their relation to other fractions of the working class. As has been seen recently in Latin American formations such as Chile, Brazil and Argentina, this 'labour aristocracy' has considerable experience in economic trade union struggle, and this has often been important for the development of socialist movements amongst the working class. Yet, at the same time, its position in the division of labour has also produced a tendency for it to attempt to maintain the differentials between itself and other fractions of the proletariat.

It seems, therefore, that the proletariat in Third World formations is differentiated into a series of clearly-demarcated fractions, and that these can be analysed as an effect of a restricted and uneven development, resulting from the particular articulation produced by imperialist penetration, in its different phases. The economic ideologies dominant in the various sectors in which these fractions exist tend, as we have briefly indicated, to be quite specific, and the effect of these on the particular forms of adherence of these fractions to political ideologies must constitute a key problem area for concrete analysis.

Capitalist penetration of agriculture and the differentiation of the peasantry

When we come to analyse the possibilities of differentiation within the agricultural sector, here much more work has been carried out, and the pitfalls to avoid are much more clearly demarcated. Analysis has gone some way beyond either a simple application of the concepts 'rich', 'middle', and 'poor' peasantry, or a dogmatic assertion that the peasantry is imbued with the ideology of private ownership and, by definition, inherently less 'revolutionary' than the small proletariat.[36] The crucial point has now been established that the investigation of the class structure of the peasantry must be based on the *degree of capitalist penetration of agriculture in the particular transitional Third World formation that is the object of analysis*. The problem that has to be posed is, therefore, why, in so many transitional formations, have capitalist relations not developed in the agricultural sector?

This cannot simply be related to imperialist strategies: whilst during the colonial phase in the formations of Africa and Asia there was no necessity for such penetration, with the development of the domestic market during the independence period, penetration of agriculture becomes a highly profitable exercise both for imperialist and national capital. Rather, as we have already stressed *the major barrier to penetration lies in the continuing reproduction of modes of production or their elements, which prevent or limit a large-scale separation of direct producers from their means of production*. We have already analysed why this is the case in—for example—the semi-feudal modes of production in Latin America, and the village units of production in the Asiatic mode of S. E. Asia. Here we can build on this by indicating how forms of class differentiation amongst the peasantry are dependent on the extent of capitalist penetration of the rural sector that has occurred in particular social formations.

Take, for example, the semi-feudal mode of production reproduced in Latin American formations. We know from our preceding analysis[37] that a capitalist penetration of agriculture is limited by the reproduction of this mode providing a material basis for the power of the landowning class in the state apparatus. This limits the differentiation of the peasantry along capitalist lines, although the demarcations that do result are quite complex, since the possible combinations of labour—as we have already noted[38]—are multiple. Such is the product of an economic structure which restricts capital penetration by retaining mechanisms preventing a large-scale emergence of wage-labour. As a result, the peasantry comes to be occupationally divided into a whole series of groupings whose differences are minimal; even the demarcations between such disparate categories as 'landholder' and 'wage-labourer' are confused by the fact that the former may have to work on a hacienda to substantiate his income, and the latter may have a minuscule plot of land on which to cultivate a small part of his subsistence crop. The existence of clearly demarcated classes is constantly blocked by the continuing reproduction of the semi-feudal mode.

A contrasting case of capitalist penetration is provided by the example of French colonial penetration of village production in N. Vietnam (Tonkin). Here, colonial policies, inserting themselves into the dynamics of an Asiatic mode which allowed temporary forms of landownership and separation of labour from its means of production, were able to produce a degree of land concentration and capital accumulation, and to create a pool of wage-labour for agriculture. Consequently, classes of a specifically capitalist form were able to emerge in the agricultural sector during the twentieth century, prior to the independence period. The colonial fervour to create a labour-force for the lucrative raw materials sector, combined with a division of labour which could provide a limited labour force, making differentiation possible, were the key determinants of the emergence of capitalist production, which then co-existed with communal village production. Thus, the division of labour in the previously dominant Asiatic mode, as it reproduced itself in the transitional formation, permitted a limited development of capitalist production, which assisted imperialist penetration in producing a capitalist differentiation of the peasantry.

From these examples, we can see that any investigation of the economic and political ideologies within 'the peasantry' must begin with an analysis of the process of differentiation, of the extent to which the continuing reproduction of non-capitalist modes or their elements can effectively block imperialist penetration of the agricultural sector. Within the same social formation, ideological forms prevalent in a familial based domestic agricultural unit whose members work for wages on a nearby hacienda will be totally different from those of wage-workers on export estates; their economic ideologies are grounded in widely disparate forms of production. Unless the problem of analysing the differing contents of economic ideologies generated in this way, and their effects on the political ideologies dominant within the various groupings of the peasantry, is rigorously posed, there can be no adequate basis for analysing the changing political alliances of the peasantry in particular social formations.

This point cannot be carried any further here. Having presented a general framework in which the process of differentiation of the peasantry can be analysed, it now seems possible to examine the class structure of the agricultural sector avoiding the method of category application in which the problem has usually been posed.[39]

The petty-bourgeoisie in Third World formations

Throughout his analysis of the capitalist mode of production, Marx had a specific notion of the petty-bourgeoisie, defining it solely as small capitalists in the retailing sector. As against this, the Marxist tradition, faced with a growth of disparate social groups who cannot be easily categorised in Marxist terms, has attempted to insert them into Marx's category of the petty-bourgeoisie, thereby producing a categorisation whose explanatory adequacy is absolutely minimal. Confronted with this, a number of authors have recently tried to specify both the limits of this class, and its various internal groupings.

The most successful attempt so far has been made by the authors Baudelot and Establet,[40] who draw an initial distinction between three fractions of this class: those who effectively control the administrative apparatuses of the state (the legal apparatus, the educational apparatus, etc.), those who own small units of production, distribution and exchange of commodities, and those who own and/or control units providing services for the production and distribution of commodities (from advertising firms to machine maintenance). Several criticisms can be made of this demarcation, but for our analysis the most relevant is that their definition ultimately assumes a political or economic cohesiveness between these various fractions. Such a conclusion may be valid for industrial capitalist formations, but it seems untenable for Third World formations, since the ultimate unity posed by Baudelot and Establet between these constitutive fractions is less evident than their constant opposition, economically and politically.

Consequently, in examining the petty-bourgeoisie here I have restricted the category to Marx's narrow definition, defining it as a group whose existence is dependent upon the place it occupies in the distribution of commodities—either produced in the Third World formation or imported into it. The stratum of state functionaries is analysed as a distinct entity, since—as I hope to show—this is more appropriate for our analysis of the class structure. Similarly, the groups of capitalist owners of small units of production in the industrial or agricultural sector can be more statisfactorily analysed by categorising them as a fraction of the capitalist class, as I indicate below. Having stated this, we can now move on to our investigation of the petty-bourgeoisie.

We should note first that the distributive function for this class is much more extensive, *vis-à-vis* the productive sector, than is the case in advanced capitalist modes of production. This is due primarily to the general absence of capitalist penetration of agriculture in the Third World formation, which enables an extensive network of trading links to be set up for commodity exchange between the capitalist and non-capitalist sectors.[41] Commodities travel along a chain that links together a series of distributors, each of whom appropriates a part of the surplus labour or surplus value already realised. This appropriation, dependent on the circulation of commodities set up by restricted and uneven development, forms the material basis for the mercantile fraction of the petty-bourgeoisie. Consequently, the perpetuation of this fraction in its 'inflated' form in Third World formations is totally bound up with the maintenance of restricted and uneven development. Such phenomena as extensive capitalist developments in agriculture, increasing state intervention to reduce value appropriated in circulation, or nationalisations of foreign or domestically owned means of production, are generally opposed by it, and, throughout the post-war period, this has been constantly reflected in its political ideologies and participation in political movements.[42]

An important fraction of the petty-bourgeoisie involved in the distribution of the surplus product consists of those who form the final link in the distribution chain, the retailing groups in urban areas with their access to the cheap labour supplied by the semi-proletariat. For this fraction, their degree of support for restricted development is often dependent on the particular commodity they trade in. For example, groups who depend on the sale of imported commodities from industrial capitalist formations, or those who trade in department 11a goods, will favour the perpetuation of this form of development, whereas those who sell domestically produced agricultural commodities will tend to support the extension of capitalist relations of production in agriculture, since this process will initiate a reduction in the power of the rural-based mercantile fraction and an overall increase in capital-intensitivity, thereby potentially reducing the price of agricultural commodities.

Within the petty-bourgeoisie, there are, therefore, differences over the degree to which the form of capitalist development presently attained in most contemporary Third World formations should be extended.

Whilst both the urban and rural petty-bourgeoisie in the distributive sector tend to be represented politically by parties committed to development strategies tied to imperialist penetration, the differing degree of support for an extension of capitalist relations can lead to potential political conflicts within this class, with rural groups favouring parties that seek to limit this extension, and some urban sections allying with those classes and class fractions that are developing strategies to extend it into the agricultural sector.

Contradictions within the capitalist class

In many analyses of Third World formations, it has been assumed that there is a major split within the capitalist class, between clearly demarcated entities termed 'national' and 'comprador' classes. Thus, for example, the strategies of many mass-based left-wing parties, particularly in Latin American formations, have been based on a support for the national capitalist class, on the basis that its conflict with the pro-imperialist comprador class renders it a progressive force. In analyses of particular social formations, there has been a tacit assumption that a potentially large progressive national capitalist class does exist, followed by a furtive search for evidence to support this assertion—evidence which often does not substantiate the premise.[43]

Rather than beginning with such an assumption, it seems more appropriate initially to locate the existence and further possible tendential development of this class within the process of restricted and uneven development promoted by the articulation we have outlined. This, of course, will vary considerably from one particular formation to another. Consequently, what we will present here is a brief outline of *the questions that should be posed* in this area of investigation.

In Third World formations, any analysis of the conditions for the existence of *a national capitalist class* must initially be related to the degree of emergence of capitalist production in those sectors not dominated or controlled by imperialism during the colonial period. Questions such as the following should be posed: To what extent does artisan production develop into the form of simple co-operation on a capitalist basis during this period? To what extent is it able to develop the productive forces beyond this point in the face of competition from imported capitalist commodities? To what degree is a market constituted for its products beyond the non-capitalist division of labour? To what extent was the position of this class strengthened by periods of decline in imperialist penetration? To answer each of these questions—which are crucial for analysing the emergence of a national capitalist class as a specific entity—we must, therefore, investigate the extent of capitalist penetration and changes in its pattern, as structured by the development of the industrial capitalist mode, and the degree of restriction imposed by the continuing reproduction of the non-capitalist mode of production, viz., the particular form taken by the articulation of modes or their elements during the colonial period.

During the post-colonial period, as we have indicated, the form of restricted and uneven capitalist development imposed on Third World formations forces indigenous capitalist units (producing mostly for the local market but also for export) into the least profitable sectors, where the development of the productive forces necessarily remains at a low level.[44] Faced with such a situation, and, often with the continuation of competition from imported commodities, the national capitalist class can either remain in these sectors, or enter those dominated by foreign and/or comprador capital. In the case of the latter, its role tends to become similar to that of comprador capital, increasingly dependent on the maintenance of imperialist penetration for its reproduction.

The major question for any analysis of a national capitalist class is, therefore: with the extension of capitalist relations into the industrial sector beyond the confines of the export sector and department 11a production, is the basis for the existence of a national capitalist class, whose form of production can be initiated and reproduced without the presence of imperialism, being eroded?

Again, this question can only be answered with regard to particular formations, when one has analysed the extent to which the restricted and uneven development promoted by imperialism can, in specific conjunctures, be extended into department 11b production, and into the agricultural sector. The main areas for investigation then become the strategic objectives of the classes and fractions dominant within the state, the profitability of this extension for industrial capitalist firms, and the solidity of the ties between domestic production and the non-capitalist divisions of labour, and so on.

In addition to these problems of analysing the conditions of existence and the changing role of national capital in Third World formations, the assertion that, *as a class*, it can always be presented as a clearly demarcated entity, must also be questioned.

In Latin American formations such as Chile, for example, the capital invested in the domestic sector often has its origins in the agricultural sector, from the accumulated surplus labour realised by the landowning class.[45] Similarly, in many African formations during the post-independence period, investment in the non-imperialist dominated sector is often based (as we have noted earlier) on the monetarised accumulated wealth of the tribal chiefs, who are also involved in comprador functions with foreign capital.

In such formations, where the previously dominant non-capitalist relations of production still persist, or where they are realised as a relation of distribution on the surplus value produced in the capitalist sector, it is hardly surprising that the functions of the national capitalist class are, to an extent, fulfilled by members of previously dominant classes. This fact has, however, been grossly neglected by most proponents of the national capitalist thesis.

The question of the existence of a national capitalist class as a clearly demarcated entity in Third World formations, is not, therefore, one that is easily resolvable. If, however, it can be specifically located in the economic articulation we have outlined, and if there is a basis in a particular formation for its perpetuation, there can be no doubt concerning the nature of its contradiction with the other major fraction of the capitalist class. Despite a basis for political unity between it and the comprador fraction against, for example, the strategies of a semi-feudal landowning class, or against any left-wing movements amongst the proletariat and peasantry, there is a permanent source of conflict between it and the comprador fraction of the capitalist class.

The *comprador fraction* is intimately tied to imperialist penetration. It can be dependent on foreign capital through its control over units of production tied by linkages to the extractive industries, or through the production of luxury commodities that depend on the extremely unequal distribution of income perpetuated by imperialism. At a more specific level, its material basis may be formed by a joint investment with foreign capital in units producing for export, or by ownership of units in import substituting industries dependent on imported capital and commodities. Its most parasitic section is what we can term the comprador-financier fraction, which has its material basis in the accumulations of banking capital realised in the comprador and foreign capital sectors; these are then generally invested in more profitable ventures outside the country, usually in the industrial capitalist economies.

We can thus see that the comprador and national capitalist fractions of the capitalist class occupy very different places in the economic structure of Third World formations. This is clearly reflected in their differing political ideologies and by the economic strategies put forward by their political representatives in the state. On the one hand, the representatives of national capital stress the need for limited protection of indigenous industries, restrictions on imported commodities, the dangers of a heavy reliance on imported technology, national control over any developments of capitalism in agriculture following a land reform, and so on. On the other hand, representatives of comprador interests tend rather to argue for the perpetuation and extension of imperialist penetration, emphasising the importance to economic development of the value accruing to the state from foreign investment in the export sector, the importance of linkage industries, the increases in profitability from the use of imported technology, etc.

There is, therefore, a major contradiction within the capitalist class in Third World formations which, as we indicate below, emerges in a political form in particular conjunctures. If the form taken by restricted and uneven capitalist development in a Third World formation enables units of capitalist production increasingly to emerge and be reproduced without the intervention of imperialism being necessary for this reproduction, and if the future possible tendential development of the economic articulation is likely to perpetuate this reproduction, then the existence of this contradiction, and its possible political representations becomes of considerable importance for concrete analysis.

The Emergence of a Stratum of State Functionaries

No matter what the development strategy pursued by a particular class or alliance of classes holding state power in whatever institutionalised form—whether it be an alliance of the national capitalist class with sections of the proletariat and peasantry pursing a form of independent capitalist development, or an alliance of the semi-feudal landowning and comprador classes following a path of restricted development in the industrial sector—it has been the case that, during the independence period, the implementation of these strategies has required a massive expansion of the administrative apparatus of the state. Whether the object of economic planning is variously to promote joint state foreign capital investment, to institute the process of import-substitution, to implement capitalist agrarian reforms, to organise the distribution of commodities, or to restrict the influence of imperialist penetration, all these have needed wide-scale state intervention within the economic structure. It is primarily this necessity, rather than any inheritance of a colonial administration, that has produced a new stratum of administrative functionaries in Third World formations. In many countries, the size of this stratum is substantial, since, in addition to economic planning, it is also required to implement the many other forms of social engineering that are prescribed by modernisation theorists as prerequisites for capitalist development.[46]

However, in addition to this, in many Third World formations the administrative apparatus of the state is inflated beyond the levels required for the implementation of development planning. Here it seems that the rapid increase in the number of administrative functionaries during the independence period must be related to the growth of the semi-proletariat in urban areas. In a country such as Indonesia, for example, where unemployment in the 1970s has been as high as 15 million, there can be little doubt that the inflation of state employment is directly related to this phenomena. Other examples, such as post-independence India and the large state bureaucracies of African formations, only reinforce this conclusion. Urban labour is recruited for menial white-collar occupations in an attempt to gain its support for the political status quo.

The increase in the number of functionaries recruited to the state apparatus must also be related to the development of the educational system that has occurred in many Third World formations,[47] particularly with the extension of capitalist relations of production beyond the traditional export sectors and the limited area of department 11a to sectors of department 11b. The required increase in the number of skilled and semi-skilled workers, which we outlined earlier in this chapter, has been met by a massive expansion of those sectors of the educational system which train these workers economically and ideologically for their place in the newly emerging capitalist division of labour in manufacturing industry.

Before dealing with the political effects of this increase in the number of state functionaries, we should briefly note, as we have done with other classes and class fractions, that not only are its determinants a general effect of the economic articulation outlined above, but, more particularly, of changes within this articulation during the post-war period.

The recent extension of department II manufacturing industry marks, for example, a new phase in the development of this articulation, evidenced by an intensification of attempts to promote capitalist relations of production in agriculture and to extend the domestic market for manufactured commodities, and by foreign capital increasingly utilising the cheap labour displaced from the countryside to produce selected labour-intensive based commodities (both for the domestic and external market) under relatively more favourable economic conditions than it can presently attain in the industrial capitalist formations. This intensification of restricted and uneven development requires an extension of state intervention (to extend the market, promote investment in specific areas, etc.) and an increase in trained labour-power; amongst its effects will be an increase in rural-urban migration, a continuous renewal of the semi-proletariat. As a result of the extension of restricted and uneven development in the manufacturing sector, we have, therefore, the conditions both for a perpetuation of the stratum of state functionaries, and an increased recruitment to its ranks.

The extension of this stratum has important political implications. It creates an expanding group which is directly dependent on the class or class alliance that occupies state power, since the strategies pursued by the political representatives of different classes will have varying effects on the level of its employment, since they can produce a relative increase or decrease in its determinants. Consequently, one of the most important political tendencies within this stratum is to favour the political dominance of alliances such as those between the representatives of the national and comprador classes, which will perpetuate restricted capitalist development and extend capitalist production into non-capitalist sectors. Yet, whilst there is a marked tendency for this stratum to act as ideological supports for such alliances, its political actions are often much less explicit, reflecting the fact that its politico-ideological formation is also determined by sections of it retaining links with its communities of origin in the non-capitalist division of labour, by its continuing contact with unemployed semi-proletarian groups, and by its exposure to the theories of socialism, communism, and Third Worldist forms of nationalism during its period of training in the educational apparatus. Consequently—and perhaps to a greater degree than those classes and class fractions we have already outlined—the political movements of this stratum are, in general, much more volatile, much less analytically predictable.

The reproduction of previously dominant classes in the new class structure

(i) The reproduction of a relation of production: the semi-feudal landowning class. We have already outlined how the continuing reproduction of a semi-feudal mode of production, and the occupation of state power by the landowning class in alliance with other classes, can be analysed as an effect of early imperialist penetration and the forms of penetration that preceded it. This position is perpetuated during the subsequent imperialist period by the alliances this class forges with imperialism and its class supports, the fractions of the comprador class, which emerge from the class structure produced by restricted and uneven development. These political alliances are initially required by imperialism for its entry into the extractive sector and then the domestic market, and by the comprador class, as a political support for its development strategies. Eventually, however, these alliances fetter capitalist production. Once the extension of department II (and particularly IIa) production has reached the limits of the internal market (which is restricted by the subsistence consumption of the peasantry and

most sections of the proletariat), the further development of capitalist production is blocked by barriers against a development of capitalism in agriculture and a subsequent extension of the internal market, both of which are the result of the continuing reproduction of the semi-feudal mode of production in the countryside.

The landowning class is fully prepared to support the development of capitalism internally provided it does not exceed these boundaries. When—despite the implementation of intervening strategies such as dependent manufacturing industry producing for export, industry assembling parts of commodities for sale in industrial capitalist formations, and so on—the limits of the internal market are reached, and when the extension of department II production and a consequent penetration of agriculture are required by both imperialism and comprador capital, the landowning class no longer supports such an extension. It then becomes necessary for imperialism and its supports to develop alliances with other sections of the capitalist class, the petty-bourgeoise, and sections of the proletariat and peasantry, to restrict the political power of the landowning class as a prerequisite for beginning to transform its economic basis. Only by gaining the support of these classes for a programme of capitalist land reform can the pro-imperialist fraction of the capitalist class succeed in extending restricted and uneven development. Yet, as we have outlined above, there are major contradictions between these classes and the capitalist class, which permanently threaten any such alliances, and if, at any moment, these begin to be manifest politically, both the fractions of the capitalist class and the political and economic representatives of imperialism retreat back into the former alliance with the landowning class, in order to preserve the existing system of restricted development which maintains them.

The political prominence of the landowning class, produced by merchant-capitalist penetration and reinforced by later forms of penetration, is thus alternately perpetuated and undercut by imperialism and its class supports. The alternation is a result of changes in the restricted and uneven development produced by the articulation of modes of production and their elements introduced by imperialism—changes which, as we have seen, are necessarily required both by imperialism and by the indigenous capitalist classes. A major contradiction between *economic requirements* and *political possibilities* is thus produced by the articulation developed by imperialist penetration. This contradiction will continue to provide the basis for further political conflicts in Latin American formations, as the restricted and uneven development initiated by imperialist penetration requires to be increasingly extended.

(ii) Political control through the representation of non-capitalist ideologies as an effect of the articulation. In the case of the landowning class in Latin American formations which we have just examined, the political position of this class is a result of the continuing reproduction of a non-capitalist relation of production in a transitional formation. Its political power is retained because the classes dominant within the capitalist mode of production are unable to break down the reproduction of a mode whose dynamics provide an inadequate basis for the emergence of the prerequisites for capitalist production. Consequently, we are faced with an interpenetration of class structures required by two very different modes of production.

What, however, of Third World formations where the reproduction of the non-capitalist mode has been broken down, where only some of its elements continue to be reproduced as a division of labour in the agricultural sector? How is the political power of the previously dominant class maintained here?

If we take as our example the case of formations previously dominated by the lineage mode of production, where the division of labour existing in this mode continues to be reproduced, we know, from our earlier analysis, that the former relation of production between elders and village units takes the form of a relation of distribution, through which the elders have access to the value produced in the capitalist mode of production. As we have seen, this process provides the elders of the tribes with a continual supply of monetary wealth, which, combined with the monetarisation of the wealth (accumulations of 'élite goods') they have amassed from their place in the lineage mode, enables them to participate in the various sectors of restricted development, fulfilling the functions of comprador capital, and, during a later phase, promoting the extension of manufacturing industry.

The existence of this material basis, together with the political dominance they retain during the colonial period through the direct assistance of colonial policies, goes some way to explaining why the class of elders is able to secure state power during the independence period; in itself, however, it is inadequate, since it omits a crucial issue—namely the way in which the political domination of this class continues to be legitimised through the continuing reproduction of elements of the ideologies of kinship existing in the lineage mode. If we take, for example, the political ideologies through which the actions of the state and the relationship between the classes and class fractions dominant and subordinate are legitimised in Third World formations previously dominated by the lineage mode, it seems that they combine elements of kinship and religious ideologies characteristic of the lineage mode with political ideologies characteristic of capitalist formations.[48]

Whilst imperialist penetration attempted, through the colonial state, to ensure the political dominance of classes that were supports for the form of economic development it required—state functionaries, the commercial petty-bourgeoisie, tribal elders fulfilling comprador functions—the independence period has not produced a stable dominance of these classes. Rather, *conflict between fractions within these classes* has continually characterised the political regimes of West-Central African formations during the post-war period. Furthermore, within the alliance colonialism attempted to create, a series of class fractions have emerged, apparently produced in the realm of political action, with no foundation in the economic structure that ultimately determines the overall economic dominance of the classes within the alliance. Consequently, the 'reality' of politics in the region appears to have been more the result of tribal, ethnic, religious and regional differences than of any oppositions created by the economic structure. Hence it seems that the successes of leaders or parties have simply been based on pragmatically playing off one fraction against another. Yet, such a conclusion is illusory. Whilst these demarcations have been important, they have operated within the limits of the class alliances required by imperialist penetration. Within these limits, many different political alliances are possible, precisely because ideologies based on these demarcations continue to be reproduced, providing a basis for political fractionalisation. Thus, the requirements of imperialism for a stable political alliance providing a guarantee for restricted and uneven development have not been met in the region in the long term, as a result of political control still being able to be legitimised through a combination of ideologies in which values and concepts prevalent within the pre-colonial formation still play a vital political role.

This phenomenon, which is of crucial importance for any concrete analysis of the formations of West-Central Africa, can be shown to result from the continuing articulation of a non-capitalist division of labour with a capitalist mode of production, which provides a constant basis for the insertion of non-capitalist ideologies within the political forms produced by a specifically capitalist class structure (parliamentary democracy, Bonapartism, fascism, etc.).

As we noted earlier, the separation of direct producers enforced by the colonial state means that the dowry required by the elder can now be obtained outside the confines of village production. With the latter being increasingly confined to smaller areas during colonialism, the supply of women and slaves can no longer be ensured by the greatly reduced surplus labour produced in the village. Consequently, it is ensured by a value (capitalist wages) obtained elsewhere. Village production still requires a supply of women for its reproduction, and the means for obtaining this supply now comes mainly from the capitalist mode. Thus, marriages continue to be arranged through the provision of dowries, and it is this material basis which is the *support for the maintenance of a kinship ideology through which the political dominance of the elders continues to be legitimised.* To the extent that the articulation between a capitalist mode of production and a division of labour characteristic of the lineage mode is perpetuated, to the extent that the capitalist mode does not begin to totally break down subsistence production in the village units, the basis for the reproduction of kinship ideologies in the state apparatus will be maintained. Since the political dominance of the elders as a class depends to a considerable extent on the continuing reproduction of village subsistence production, it is likely that, despite their comprador function, they will resist any rapid process of capitalist penetration that produces a differentiation of the peasantry, in favour of a more prolonged period of combination of capitalist and non-capitalist production in the villages.

From this example, we can see that even where imperialist penetration breaks down the reproduction of the non-capitalist mode, the political position of the previously dominant class can still be perpetuated through the maintenance of the non-capitalist relations of production as a relation of distribution on the value produced in the capitalist mode. Furthermore, as with the transformation of a feudal relation of production—absolute ground-rent—into a relation of distribution in the European capitalist formations, so with formations previously dominated by the lineage mode of production, it seems that this material basis for the political prominence of the previously dominant classes can only be undermined by the political action of the representatives of those classes specifically required for the reproduction of the capitalist mode of production, in a situation where those holding political power in the state apparatus are not concomitantly involved in both capitalist and non-capitalist forms of production, and where the extension of capitalist relations of production is required by imperialist penetration.

In this section, we have outlined the major classes and fractions produced in Third World formations *as an effect of the articulation of modes of production and divisions of labour resulting from imperialist penetration.* Imperialism intervenes in this class structure to promote the dominance of a political bloc within the state apparatus which will ensure the reproduction of its penetration internally, and *establish the basis for the form of post-independence political regime.* In the following section, we will outline how two different forms of these regimes can be analysed *as variants of the general class structure we have outlined above, variants that are dependent upon the specific characteristics assumed by the economic articulation and its restricted and uneven development in different formations.*

Example I: the political dominance of the comprador capitalist and landowning classes. Despite the problems that can emerge over the extension of capitalist relations of production into the agricultural sector, initially the most favourable political alliance for imperialist penetration is that between the fractions of the comprador class and a landowning or formerly dominant non-capitalist class integrated into the non-capitalist mode, whose main supports are sections of the petty-bourgeoisie and a relatively large stratum of state functionaries. The national capitalist class is neutralised by various means, and since the strategy adopted depends upon the use of cheap labour in industry and results in the neglect of domestic agricultural production, the proletariat, peasantry and semi-proletariat are necessarily subject to extensive forms of political oppression.

The strategy pursued by such an alliance of political classes depends heavily on the utmost extension of restricted development in the industrial sector. The surplus value realised in the agricultural export or extractive sectors is largely appropriated by imperialism. What remains accrues to the comprador class who—often via the state—invest it in industries tied to the extractive sector (processing, etc.) or dependent upon imported technology. The agricultural sector, controlled by the landowning or previously dominant non-capitalist class, is geared to the reproductive needs of the industrial capitalist mode, and production for the domestic market is kept to be an absolute minimum. The subsequent need for an extensive distribution network to ensure the circulation of agricultural commodities, combined with the distributive requirements of commodities imported from other capitalist modes considerably strengthens the other major class support of this strategy—the mercantile petty-bourgeoisie. The domestic agricultural sector is largely left to stagnate, whilst the policies pursued are detrimental for the national capitalist class. Consequently, the semi-proletariat tends to increase, due to capital-intensive manufacturing production and the decline in subsistence agriculture. The strategy, in general, provides a basis for political conflict between its dominant classes and, particularly, the peasantry, semi-proletariat and national capitalist class. Because of this, and also because, at a later stage in its development, it may become necessary to reduce the political prominence of the landowning class in order to initiate a capitalist penetration of agriculture, this alliance often exists in the form of a military domination. When we examine contemporary countries such as Brazil, Chile, and—our example below—Indonesia, this becomes clear.

If we now take the case of Indonesia, we can briefly indicate how the emergence of this alliance can be analysed as an effect of a particular articulation of modes of production and divisions of labour produced by imperialist penetration; how the form of restricted and uneven development subsequently produced by this articulation led to a specific political polarisation that resulted in the political dominance of this alliance.

As we have already noted, through its long period of penetration under the dominance of merchant's capital and its extensive use of enforced cultivation during the commodity export phase, Dutch colonialism reinforced the existing non-separation of direct producers in the agricultural sector. The general confinement of imperialist penetration to the raw material sector and to a relatively limited plantation sector,[49] combined with the continuation of commodity export for an agricultural sector based on small peasant farms and plots of land, enabled the non-capitalist division of labour, already reinforced by previous penetration, to be strengthened under the control of the large village landowners. This particular confinement of development produced by Dutch colonialism is a crucial factor in explaining the continuing dominance of a landowning class within the economic structure.[50] The fact that penetration was confined to these sectors also established a basis for a fairly cohesive comprador class which, however, did not really emerge until after independence.[51] Despite the perpetual opening up of the domestic market to imported goods, a small national capitalist class was able to develop in the domestic sector emanating specifically from artisan production in the villages.[52] The mercantile petty-bourgeoisie[53] was strengthened considerably during the colonial period, due to the openness of the domestic market, the purchase and sale of commodities produced on millions of small peasant allotments controlled by the landowning class, and so on. The proletariat remained necessarily small,[54] whilst—as a result of the decline in subsistence production rather than through direct colonial intervention—the semi-proletariat increased,[55] particularly as the landowning class gained greater economic control in the countryside.

As a result of this particular articulation, political conflicts in Indonesia during the independence period can be analysed as centring around a specific opposition of classes, induced by colonialism yet subject to minor modifications resulting from changes in the determinants of this articulation during the post-colonial period.

The general opposition combines the support classes for Dutch colonial penetration—the village landowning, comprador and mercantile petty-bourgeois classes—seeking to extend imperialist penetration within the industrial sector, against those who are most thoroughly exploited in this specific form of restricted development, namely the peasantry and the semi-proletariat. The national capitalist class tended, for reasons outlined below, to oscillate between support for particular classes on each side of the opposition. The proletariat, remaining relatively small, generally allied itself with the semi-proletariat and peasantry—despite the fact that many sections of the latter have articulated their opposition to the dominant classes in ideologies that combine those of the urban petty-bourgeoisie and national capitalist class[56] with their own animist ideologies[57] to produce brands of populism that are hostile to all classes involved in capitalist production.

If we now take one particular period of political conflict in Indonesia—that of the 'Guided Democracy' period from 1957–65[58]—we can analyse it as a particular variant of this opposition induced by attempts to transform the articulation on which it rests.

The national capitalist class, limited by the extension of restricted development in the industrial sector during the post-1949 period, saw a possibility for extending the market through the limited proposals for capitalist land reform put forward under Sukarno's Bonapartist regime in the early sixties.[59] This move against the landowning class was initially tacitly supported both by imperialism and, hesitatingly, by the petty-bourgeois classes against opposition from most sectors of the comprador class.[60] This split in the colonial alliance, induced by the extension of restricted development beyond the extractive sector in the Guided Democracy period, was accompanied by the development of a populist ideology,[61] particularly amongst the semi-proletariat and peasantry, which tried to unite these classes with the national capitalist class, stressing the centrality of individual ownership of land combined with a notion of village communalism, directed against the landowning class.

In this favourable conjuncture, the national capitalist class attempted to break further out of its confinement by arguing for the protection of the domestic market from imported commodities and the implementation of nationalisation in sectors dominated by foreign capital.[62] This strategy, seeking to limit restricted and uneven development not by confining its extension, but by undermining it, began to re-unify the colonial alliance, with the full support of imperialist penetration. Utilising the anti-capitalist elements of the populist ideologies amongst the peasantry, the colonial alliance itself now attempted to break up the opposition bloc. Thus it was able to achieve, culminating in the brutal military coup in 1965, which re-established the political dominance of the colonial alliance, this time in a military form.[63]

Specific changes in the nature of the articulation of a capitalist mode of production with a non-capitalist division of labour—the extension of restricted and uneven development in the industrial sector and the need for imperialist penetration to gain access to the agricultural sector—resulted here in attempts to transform the political opposition of classes produced by imperialist penetration in its colonial period. Yet, once the restricted and uneven development that this alliance produced was itself brought into question, then the alliance of classes acting as supports for its development was able to re-establish its political domination.

The 'Guided Democracy' period in Indonesia can thus be seen as centring around a class conflict induced by a particular variant of the general opposition and alliances of classes produced by an articulation specific to the Indonesian formation as a result of capitalist penetration within it. The operation of this variant is limited precisely by the confines of the general class structure in which the articulation forces it to operate.

Example II: the political dominance of fractions of the comprador class and the national capitalist class. When a political alliance emerges between fractions of the comprador class and the national capitalist class, and this alliance is able to secure state power, this provides a means for imperialism to begin to penetrate the agricultural sector of Third World formations, by supporting an alliance that can begin to limit the political power of the representatives of the landowning class in the state apparatus. Such an alliance, however—whose material basis becomes increasingly important when uneven and restricted development has extended into department II production, and the penetration of agriculture becomes essential for the national capitalist class and beneficial for imperialism—tends to be unstable. Attempting to make the peasantry and the various sections of the proletariat cohere into a political bloc against the landowning class and its major support, the rural-petty-bourgeoisie, creates a series of major political problems, particularly where the alliance exists in a Third World formation with an articulation of *modes of production.*

This can be illustrated by referring to a specific example of this alliance, as it emerged during the period 1964–70 in Chile under the presidency of Eduardo Frei.[64]

The main period of industrialisation in Chile prior to 1945 had been organised from above during the 1930s and 40s. It had been based on import-substitution promoted through state intervention in the economy, assisted by the temporary limits placed on imperialist penetration by depression and war.[65] By the end of the war, this industrialisation began to decline. Between 1930 and 1950, for example, 149 new firms had been established, whilst between 1950 and 1960, only 21 were established.[66] The decline was primarily the result of the saturation of the internal market for consumer goods, largely amongst the capitalist and landowning classes. The perpetuation of low wages and the maintenance of an unproductive and backward agricultural sector blocked any further extension of the market. Whilst the landowning class was quite prepared to invest in a limited capitalist development by supporting medium and large-scale industries, it was unwilling to extend this beyond limits that would threaten its control over the agricultural sector, which continued to be geared towards the production of commodities for export. Consequently, the industrialisation of the 1930s and 40s, which had been based on a growing inter-relation between comprador and landowning capital (revealed in the fact that almost one half of the large businessmen in Chile own big farms or are closely related to owners of them[67] was slowed down by what has been aptly termed by Petras,

> . . . the industrialists' inability to overcome their dual role: commitment to industrialisation on the one hand and to a narrow internal market and limited effective demand maintained by low wages and an unproductive system of land tenure of which they are a part on the other.[68]

Thus, the extension of restricted and uneven development was blocked by the continuing reproduction of a semi-feudal mode of production, the surplus product of which founded a limited growth of comprador capital, but which was ultimately detrimental to the capitalist class as a whole—and particularly to what we can term the national capitalist class in Chile, who, producing for the domestic market in small units of production,[69] constantly come up against the barrier of limited demand.

Following the decline of state-induced industrialisation, the economic policies pursued by regimes in Chile during the period 1952–64 can be seen as attempts to handle problems generated by the limits placed on the extension of an economic development based on the domestic market. Thus, from 1952–8, the Ibanez regime, based on a populist alliance against the landowning-comprador oligarchy, attempted to cope with these problems by curbing widespread inflation resulting from the decline of the industrialisation period, fuelled by a fall in copper revenues. The barriers this placed on industrial growth and the oppressive policies required to hold down wages brought resistance from both the oligarchy and the peasantry and working class, and an end to Ibanez's presidency by 1958. This was then followed by an attempt to promote economic development, this time remaining within the confines of the existing articulation of modes of production.

Alessandri, a leading industrialist and son of one of Chile's large landowning families, promoted a policy of 'restoring free competition' in the private sector, and reducing state intervention. The subsequent inflation this produced could only be prevented by a massive entry of foreign capital and aid to expand state sector expenditure (primarily infrastructural development). This led to a rapid increase in foreign debt,[70] to a strengthening of foreign capital in the industrial sector, and to an increasing dependence on foreign loans and capital equipment for industries in the domestic sector. Alessandri's attempts to reduce reliance on foreign aid by 'controlling inflation'[71], through holding down wages, coupled with the unemployment effects of small industrial firms going bankrupt led to a massive alienation of the working class. Meanwhile, the effects of increasing dependency on foreign capital were proving disastrous for the small domestic capitalist class, and sections of large-scale capital began to suffer from the cut-backs required to reduce foreign debts. Consequently, the demise of Alessandri enabled a new alliance, based on the temporary mutual interests of fractions of the comprador class and national capitalist class, and drawing support from the urban white collar stratum, to gain power in 1964 under Eduardo Frei.

Frei's Christian-Democratic Government was based on an alliance between the fraction of the capitalist class not within the industrial-cum-landowning group, the Chilean white-collar stratum, and sections of the working class and peasantry. Its accession to power was greatly facilitated by the U. S. Government which, during the Kennedy administration, was promoting the new 'Alliance for Progress', seeking a reforming capitalist alternative to its traditional Conservative support, since, for capitalist penetration, it was becoming essential to open up the internal market in Latin American countries, providing capital equipment for agriculture and industry. In order to do this, the American government had to promote domestic industry and a capitalist land reform. Furthermore, the effects that the Cuban revolution was having on politics in Latin America were increasingly pressurising governments to grant social reforms in order to stem the growing support for revolutionary movements. The Christian-Democratic government thus derived its major support from those groups of the Chilean capitalist class who were willing to break with the capitalist-landowner alliance, in an attempt to create a basis for the expansion of domestic industrial production.

Frei's economic policies required an economic growth based primarily on the expansion of the home market, but also geared towards export. This growth would be assisted by imported capital equipment and joint programmes with foreign capital. Agrarian reform would increase the size of the home market and make Chile less dependent on food imports. The object of the agrarian reform was to create co-operatives that would become family farms, thereby creating a new capitalist class of small peasant-owners. The costs of implementing the land reform and developing domestic industry were to be met through tax reforms, the expansion and 'Chileanisation' of copper (the state purchasing 51 per cent of the shares in the copper mines), and foreign loans and aid.

Despite initial successes in increasing the output of manufacturing industries, through government expenditure and foreign aid, and a hesitant beginning to land reform, Frei's policies soon met with obstacles. The revenues obtained from the copper industry were much lower than had been initially envisaged, due to Chile's continuing dependence on external price fluctuations and Frei's 'concession' to the copper companies that they could repatriate more of their profits in return for a reduction in their shareholdings. Consequently, less revenue was generated internally for domestic industry, which thus became more dependent on foreign financing, enabling foreign firms to direct Chilean production to their own needs rather than to domestic requirements. In an attempt to create investment for domestic capital, Frei introduced a savings scheme that held down wages in the domestic sector. This produced widespread resentment amongst the working class, and increased their level of militancy during the post '67 period. At the opposite end of the social structure, the landowning class was continually blocking the implementation of the land reform. An extremely watered-down version of the original proposals finally became law in July 1967, and by 1969 only 468 properties had been expropriated, not all of which had been handed over to peasant proprietors.[72] The blocking of the land reform was having its effect on the peasantry; the number of occupations of farms increased from 16 in 1968 to 368 in 1970, prior to Frei's downfall.[73]

Thus, the domestic-based fraction of the Chilean capitalist class was increasingly faced with opposition from the peasantry, the working class, and sections of the white-collar stratum, the latter suffering from a declining standard of living and inflation in the urban areas[74] produced by Frei's policies; it was also faced with an increasing dependency on the industrial capitalist powers that was counteracting the original aim of developing industry geared towards the domestic market. In addition, the resistance of the landowning class was preventing this key element in the strategy from being realised.

The support for the class alliance on which Christian-Democratic rule rested thus began to decline; and, most dangerously for the Chilean landowning class and capitalist class as a whole, it declined to the benefit of those parties whose programmes were based on a commitment, albeit long-term, to reform agriculture and to restructure the economy in a general socialist direction.

Despite this, the domestically oriented capitalist fraction continued to support the Christian-Democratic party in the 1970 presidential election. The landowning-industrial alliance, meanwhile, fielded its own candidate, Alessandri, the leader of the Nationalist Party. The Chilean capitalist class was thus split, and this enabled the Popular Unity coalition of Socialist and Communist parties narrowly to win the election, increasing its support amongst the working class and peasantry, assisted by the adverse effects of Frei's policies.

From this particular example, we can see that the holding of state power by an alliance between fractions of the capitalist class, in opposition to a non-capitalist class whose economic dominance remains ensured through the continuing reproduction of a non-capitalist mode of production, is extremely precarious. Because the policies pursued by this alliance in relation to the agricultural sector necessarily entail conflicts both with the landowning class, and, as a result of this, with the subordinate classes in both the capitalist and non-capitalist modes, the attempt by this alliance to extend uneven and restricted development into the agricultural sector, even if it is assisted by imperialism, is extremely difficult. The main source of this difficulty appears to lie in a process of blocking by a class structure founded on an articulation of modes of production required for capitalist penetration to establish itself. Once this establishment has occurred, any extension of restricted and uneven development beyond the industrial sector—which increasingly comes to be needed by fractions of the capitalist class and beneficial for imperialism—necessarily creates conflicts between classes that threaten the very reproduction of this development itself. Such are the results of a strategy pursued by this particular alliance in a formation that remains dominated by an articulation of a capitalist with a non-capitalist mode of production.

A framework for analysing the class structure of Third World formations

In this section on the class structure of Third World formations, we have attempted to outline how the articulation of modes of production and divisions of labour produced by imperialist penetration, and the restricted and uneven development it gives rise to establish the material basis for a class structure specific to these formations. The political ideologies that emerge from the constituent ideological forms produced and combined in this structure result in varying combinations of oppositional class blocs, varying as effects of the specific form taken by the articulation in different formations.

In our examples of political alliances that emerged to occupy state power in the independence period, we have tried to briefly indicate how these can be analysed as temporary 'resolutions' of particular political oppositions, generated from class structures based on specific articulations.

THE ANALYSIS OF IDEOLOGY IN THIRD WORLD FORMATIONS: AN INITIAL APPROACH

Imperialist penetration attempts to guarantee the reproduction of its restricted and uneven development through the promotion of political alliances representing its economic class supports. Yet, this, in itself, is insufficient, since it also requires an ideological guarantee. The analysis of this problem forms the object of our final section.

The inadequacy of existing perspectives

In those texts where analysis of the ideological effects of colonialism and imperialist penetration has not been informed by the Sociology of Development—locating indices of the emergence of elements of ideologies characteristic of industrial capitalist formations—it has generally been examined either with the notions of underdevelopment or, more usually, from within the Sartrean discourse of writers such as Fanon and Césaire, presenting largely descriptive accounts of the impact of capitalist ideologies on different strata during the colonial period. Since the main aspects of this latter discourse have been demarcated and criticised elsewhere,[75] and since we have formulated a critique of both the Development and Underdevelopment approaches, it is unnecessary to examine the adequacy of these discourses for the more particular problem of analysing ideologies in Third World formations. Suffice it to say that the recent critique of approaches which have prevailed for some time in this area of research leaves us with a situation in which we are faced with a major theoretical vacuum. Consequently, what follows will necessarily be very schematic, outlining in general terms how the problem can be approached from within the framework we have outlined.

The dislocation of ideology

It is essential to begin with the dislocation between the ideological and the other levels which is characteristic of Third World formations.[76] Whilst imperialist penetration attempts to create the restricted and uneven development that its economic reproduction requires, the articulation on which this is based necessarily generates political conflicts dependent on the reproduction of class structures which are not solely capitalist. This dislocation between the political and economic levels, combining the reproduction of a capitalist mode with political alliances that block its extension, also co-exists with a further dislocation between both these levels and the ideological—resulting from the unevenness with which ideologies required for the reproduction of the capitalist mode can emerge within the transitional Third World formation. It is, therefore, this unevenness that has to be explained.

A co-existence of ideologies required for different modes of production

In our analyses of non-capitalist modes,[77] we indicated that their reproduction required the dominance of particular ideological notions, which permeated all the various ideologies existing in the social formations in which these modes are dominant. In the case of the lineage mode, for example, we saw that its reproduction required the dominance of the ideology of matrilineal kinship, embodied in ritual practices such as magic and sorcery, whilst, in the Asiatic mode, we saw that the ideology of divine kingship, exercised through the action and intervention of the state apparatus, is necessarily determinant for the reproduction of this mode's structure.

In both these modes of production, once the reproduction of their determinant relation of production is broken down by imperialist penetration, these notions and the ideologies in which they exist continue to be reproduced, as a result of their relatively autonomous character.[78]

Yet, concomitantly, imperialist penetration, both through the reproduction of its restricted and uneven development and the action of its colonial state, produces the emergence of new ideologies specifically required for the reproduction of the indigenous capitalist mode. These can emerge in three different ways. First, as specifically economic ideologies, generated as a result of the new division of labour created by restricted and uneven development. We examined the different forms taken by these in an earlier part of this chapter.

Secondly, as a result of the development of new relations of distribution and consumption produced by this division of labour, the places accorded to new classes such as the comprador class and the insertion of non-capitalist classes into dominant positions in the indigenous capitalist mode enable them to accumulate sums of capital which, given the requirements of the industrial and indigenous capitalist modes for the importation of commodities and the development of the domestic market, establish the basis for qualitatively new life-styles, based on a growing consumerism. This generates a whole new matrix of ideologies, in which the dominance and subordination of classes required by the perpetuation of restricted capitalist relations of production are expressed in the relative ability to consume luxury goods and emulate the major trends of industrial capitalist culture.

Thirdly, by contrast, we have those ideologies that are specifically introduced by imperialism either to ensure the economic reproduction of restricted and uneven development, or to guarantee it politically.

Examples here are legion: educational ideologies required for the training of subjects for places in the capitalist division of labour; political ideologies legitimising the form of parliamentary democracy, familial ideologies introduced to support migrant labour's break with the extended family, and so on.

Thus, within Third World formations, we have an increasing co-existence of ideologies required for the reproduction of both capitalist and non-capitalist modes of production. Yet—and this point is crucial—these ideologies co-exist in very different types of institutions.

The institutional framework of ideological co-existence

First, there are those non-capitalist institutions that can continue to be reproduced when the capitalist mode becomes dominant. If we take, for example, an institution that is ideologically crucial for the reproduction of the division of labour in the lineage mode of production, namely the extended family with its stress on patrilocal kinship in the lineage units, this can also be reproduced under the dominance of the restricted capitalist mode, but in a transformed ideological form. Whilst the continuing reproduction of the tribal division of labour in the subsistence agricultural sector perpetuates this ideology, the fact that individuals in this system also work as migrant labour in a capitalist mode whose increasing need for mobile labour can generate a more nuclear family form, introduces a co-existence of varying ideologies, and a potential conflict between their conflicting demands.

Similarly, if we take another ideological apparatus, the political, in, for example, Latin American formations, here we have a co-existence, inter-penetration and conflict between the political ideologies required for the representation of classes dominant in different modes of production.

Secondly, there are those institutions that are specifically required for the reproduction of the capitalist mode, which are either introduced by the colonial state or (in its absence) by the class supports for imperialist penetration. Whilst, as distinct from the institutions we have mentioned above, these are not characterised by the continuing reproduction of ideologies for two modes of production, their dominant capitalist ideologies are nevertheless necessarily penetrated by elements of non-capitalist ideologies reproduced by subjects entering from the non-capitalist division of labour. This has been important, for example, in the educational apparatus, where, particularly during the colonial period, the contradictions between the theoretical ideologies promoted by the state and the matrix of ideologies originating in the apparatus of the previously dominant mode generated political ideologies that played a crucial role in the independence movement.[79]

The determinants of ideological conflict

Within the institutions of the Third World formation, it thus seems that we are faced with a combination, inter-penetration and conflict of ideologies, produced as an effect of the economic articulation induced by imperialist penetration.

Yet, the extent of this conflict, and the sites of its occurrence seem to be dependent upon two phenomena: first, the relative degree of extension of restricted and uneven development, and secondly, changes in the political strategy of imperialism, as it attempts to ensure the reproduction of its restricted mode in different periods of its extension.

Specifying these determinants enables us to demarcate, in a general way, which will be the *primary arenas of ideological struggle during particular moments in the development of the Third World formation*. These struggles will, however, be lived in very different ways by the various classes and fractions produced by the articulation, as it develops, and they will be structured along the lines of the general oppositions and alliances of classes that we have outlined.

Developing these points will enable us to approach the crucial problem of how to produce a framework for analysing the *relative importance of different conflicts in particular institutions for the development of specific Third World formations*.

The effect of restricted and uneven development

The level of extension of restricted and uneven development within and beyond the industrial sector depends upon the reproductive requirements of imperialist penetration in its phases, and on the possibilities for the continuing reproduction of the non-capitalist system of production: The progress of this extension requires new institutions and an intensification of the insertion of capitalist ideologies into the combinations existing in non-capitalist institutions.

For example, in formations where restricted development is confined to the raw material and processing sectors, the ideological effects of penetration will pervade the political apparatus, but in other institutions, they will have a minimal influence. Conversely, if we take a Third World formation in which restricted and uneven development has extended into department II production, under the increasing political dominance of classes requiring a capitalist penetration of agriculture, then ideological inter-penetration and conflict will pervade the institutions characteristic of two modes, whilst conflicts of a specifically capitalist form will have emerged in the apparatus required for the reproduction of the capitalist mode.

If we approach this problem on a more specific level, once we have established the nature of the dominant articulation and the extent to which particular changes in the reproductive requirements of its determinants will effect its tendential development, then we can locate both the likely areas of emergence of particular forms of capitalist ideology and the areas of conflict in those institutions which increasingly require the insertion of capitalist forms. This is of crucial importance for determining the *sites of political conflict* in which the differentiated class structure created by restricted and uneven development can operate.

Changes in the political strategy of imperialism

This process of the ideological effects of an extension of restricted and uneven development is also influenced by our second determinant, above, since the extension of capitalist forms of ideology must be ensured, in each of its phases, by a political dominance of the economic class supports appropriate to these phases. Thus, during each phase of extension, the various ideologies in which these dominant classes live their relations with other classes and fractions subordinate in the class structure begin to gain ideological dominance in the major institutions of the Third World formation.

For example, in its early stages, imperialism utilises the existing ideologies of the previously dominant classes to guarantee its economic penetration. Conversely, at later stages of penetration, when the extension of restricted development requires the dominance of specifically capitalist classes, the need to retain the forms in which the authority of these classes is legitimised in the state apparatuses is no longer necessary. Consequently, whilst the extension of restricted and uneven development requires a progressive undermining of the ideologies existing in the institutions required for the reproduction of the non-capitalist mode, this process is, as it were, 'checked' by the fact that, until the moment when this extension requires the emergence of capitalist relations of production in agriculture, imperialist penetration utilises the political ideologies of the previously dominant class. Consequently, whilst the extension of restricted and uneven development does not specifically require it, the need to politically guarantee this extension nevertheless promotes the conservation of these ideologies within the non-capitalist institutions.

From this, we can see that the specification of possible sites of ideological conflict cannot simply be read off from the need to extend restricted and uneven development, as governed by changes in its determinants. Rather, *the need to politically ensure this extension generates a more complex process of ideological conservation—dissolution* in each of the institutions characteristic of the non-capitalist social formation—dissolution facilitating restricted development, and conservation ensuring its reproduction, but only up to a particular stage in its extension. Thus, whilst the extension of restricted and uneven development in the industrial sector establishes conditions that can exacerbate ideological conflicts in the non-capitalist institutions, thereby centring political oppositions around them, these conflicts cannot begin to be resolved in a capitalist direction, nor political conflicts centred on them begin to be expressed in a capitalist form, until the conditions perpetuating the insertion of the previously dominant ideologies into these institutions have been undermined. Until this moment is achieved, despite imperialist penetration's increasing economic need for a dominance of capitalist ideologies within the state apparatus, major ideological contradictions will continue to be generated in the non-capitalist institutions—contradictions which, in each phase of restricted and uneven economic extension, will provide bases around which political conflicts generated by the class structure can centre.

Having briefly demarcated the determinants of these ideological contradictions, we can now turn to the problem of their relation to the alliances and oppositions of classes generated by this class structure.

Ideology and the class structure

The political ideologies in which the oppositions between classes and fractions are expressed, appear, from the above, to be the result of an inter-penetration of ideologies dependent on a specific level of extension of restricted and uneven development. This level of development requires the political dominance of a particular bloc of classes, whose position is legitimised both by an ideological conservation—the continuing reproduction of non-capitalist ideologies—and dissolution—the emergence of ideologies specifically required for the reproduction of the capitalist mode of production—whose elements co-exist in the institutions of Third World formations to varying degrees in different phases of the extension of restricted and uneven development.

Yet, precisely because this process of conservation-dissolution varies in its effects from one class and fraction to another at different periods, the relation of these classes to the conflicts generated in the institutions of the Third World formation will be very different.

If we return to our example, above, of the co-existence of ideological forms made possible by the articulation of two modes in the familial or political apparatuses, we can see that the lived relations of different classes to the conflicts generated will be variable, depending on the level of extension of restricted and uneven development and the articulation that produces it. In the familial apparatus in the lineage mode, for example, the movement of labour from the non-capitalist division of labour to the capitalist mode initially establishes a basis for the insertion of elements of capitalist familial ideology into the reproduction of the extended familial form. The contradictory requirements for the reproduction of the two familial types come to be expressed differently within the political ideologies of the various classes and fractions existing in the non-capitalist division of labour, as a result of their degree of involvement in other institutions penetrated by capitalist ideologies. However, with such changes in the articulation as the extension of capitalist relations of production in the industrial sector and a greater degree of rural-urban migration, combined with, for example, an increasing control over state power by an alliance which is undermining the political power of those groups whose dominance is based solely on their position in the non-capitalist division of labour, conceptions of the capitalist nuclear form can now no longer be confined to classes—such as migrant labour and the semi-proletariat—participating in other, more penetrated, apparatuses, but must begin to enter as elements into the ideologies of those classes permanently located in this division of labour. Consequently, the contradiction between the nuclear and extended forms now begins to be expressed in the political ideologies of the peasantry to a greater extent than previously, when it formed only a relatively minor element, inserted through the incorporation of the ideologies of other classes involved in both the capitalist and non-capitalist sector. Here, then, as with other institutions, the ideological conflicts generated through the necessary interpenetration of ideologies are expressed differently amongst the classes and fractions produced by the economic articulation. In addition, this process changes as restricted and uneven development and its political guarantees in the class structure are themselves subjected to change.

An inter-penetration of ideologies required for two modes of production and their determinants

In this section, we have tried to outline how the unevenness with which ideologies required for the reproduction of restricted and uneven capitalist development emerge must be specifically related to the reproduction of an inter-penetration of ideologies required for two modes of production. This inter-penetration exists both in institutions characteristic of the non-capitalist formation and in institutions specifically required for the reproduction of the capitalist mode of production. We have also tried to show how this inter-penetration changes over time in particular institutions, as a result of a process of ideological conservation-dissolution, governed by two determinants.

By approaching the co-existence of different ideological forms in this way, we can demarcate both the future possible sites of ideological conflict and the forms that such a conflict will assume in particular Third World formations.

However, beyond this, we also need to know how—and to what extent—these conflicts will be incorporated in the political ideologies of the particular classes and fractions produced by the economic articulation. This problem must similarly be analysed as an effect of the level of extension of restricted and uneven development, and of the political dominance of specific class alliances during particular phases of this extension, since it is these phenomena that determine both the emergence of capitalist ideologies required for restricted and uneven development and the possibilities for the continuing reproduction of the political ideologies in the non-capitalist division of labour or mode of production.

Our analysis in this chapter has centred around the articulation of modes of production or divisions of labour brought about by the displacement of the determinant instance of the non-capitalist mode that imperialist penetration produces in Third World formations. This articulation establishes the basis for what we have termed a restricted and uneven capitalist development specific to these formations. This form of development is structured both by the reproductive requirements of the industrial capitalist mode of production and by the possibilities for the continuing reproduction of the non-capitalist mode or its elements, embodied in particular divisions of labour. These possibilities vary from one non-capitalist mode to another, and can only be rigorously approached from an analysis of the structure, reproduction, and dynamics of the particular mode that was dominant in any given social formation prior to imperialist penetration.

Similarly, changes in the form of restricted and uneven development must be analysed as the result of transformations in the determinants of the articulation of modes of production that governs it. Although we have not theorised such transformations in the way that, for example, we tried earlier to situate the emergence of imperialism as a stage in the dynamics of the capitalist mode, we have indicated how they can produce extensions of restricted and uneven development, extensions which have considerable importance both for the development of the class structure and for the ideologies dominant within the institutions of Third World formations.

In each of the sections of this chapter, we have tried to show how particular phenomena characteristic of Third World formations can be analysed as effects of this articulation of modes of production, and of changes within it. Thus in our examples of such economic features as urban unemployment, combinations of different types of labour, and accumulations of indigenous capital, we indicated how these could be analysed as forms whose determinants—the changing reproductive requirements of the industrial capitalist mode and the level of resistance of the non-capitalist mode of production or division of labour—were necessarily absent in their phenomenal appearance.

Similarly, we outlined how the economic articulation necessarily produces alliances and oppositions of classes and class fractions that are specific to Third World formations. We noted how the forms of political representation of these classes change as a result of transformations in the constituent elements of their political ideologies—transformations which, again, we have tried to analyse as effects of extensions of restricted and uneven development, which are themselves dependent on the reproductive requirements of the articulation of modes of production.

In the case of the particular forms of ideological combination and inter-penetration characteristic of Third World formations which we described above, these can again be analysed as the result of ideological conflicts produced by extensions of restricted and uneven development and the—often conflicting—need to politically ensure the reproduction of this development.

Finally, in our examples of varying development strategies promoted by alliances of classes and fractions politically dominant in specific Third World formations in particular periods of their post-war history, we briefly attempted to indicate how the theoretical framework we have outlined can be utilised to analyse the origins of such alliances, as concrete examples of particular articulations and changes within them. Our analysis has necessarily been confined to a general level of abstraction, since our object was to outline a general framework within which analyses of particular Third World formations can begin to be developed.

Our earlier critiques of the Sociologies of Development and Underdevelopment, and of recent approaches to the analysis of non-capitalist modes of production, attempted to indicate the major inadequacies of these discourses. It is hoped that our comments in this chapter, together with our analysis of the structure, reproduction and dynamics of non-capitalist modes of production, and the general effects of different forms of capitalist penetration on non-capitalist modes previously dominant in Third World formations, will stimulate some new directions for research in the analysis of particular Third World countries. If this text contributes in any way to going beyond the present situation of theoretical impasse in the analysis of those societies previously examined from within the Sociologies of Development and Underdevelopment, then it will have been worthwhile.

Notes

CHAPTER 1 THE SOCIOLOGY OF DEVELOPMENT: THEORETICAL INADEQUACIES

1. R. Dahrendorf, *Class and Class Conflict in Industrial Society*, R.K.P., London (1957). See also: 'Out of Utopia', *American Journal of Sociology*, No. 64, pp. 115–27.
2. J. Rex, *Key Problems in Sociological Theory*, R.K.P., London, 1961.
3. D. Lockwood, 'Social Integration and System Integration' in *Explorations in Social Change*, ed. G. K. Zollschan and W. Hirsch, R.K.P., London, 1964. Also, 'Some Remarks on the Social System' in *British Journal of Sociology*, Vol. 7, 2, 1957, pp. 134–46.
4. See, T. Parsons, *The Social System*, Chapter X, R.K.P., London, 1951.
5. Barrington Moore Jr., (Ed.,) 'The New Scholasticism and the Study of Politics' in *Political Power and Social Theory: Seven Studies*, Harper and Row, 1965.
6. D. Foss, 'The world view of Talcott Parsons', in M. Stein and A. Vidich, *Sociology on Trial*, Prentice-Hall, 1963.
7. C. Wright Mills, *The Sociological Imagination*, Oxford University Press, New York, 1959.
8. See our comments on Parsons's notion of 'analytical realism' (pp. 7–9) and also in T. Parsons, *The Structure of Social Action*, McGraw-Hill, New York, 1948, pp. 728–30, section 'Empiricism and Analytical Theory'.
9. In particular see, for example, the empirical essays in *Essays in Sociological Theory*, Free Press, 1954.
10. See, *The Sociological Imagination*, op. cit., 7 above.
11. Barrington Moore Jr., op. cit., 5 above.
12. Although I do not utilise the concept extensively in my text, I have used the term *discourse* here because it seems to me to be essential to view the texts of Parsonian theory as being the result of a practice of theoretical abstraction whose determinants are theoretically discernible. (The character of this process of abstraction is outlined later in the chapter.) The texts of Parsonian structural-functionalism are, therefore, viewed as component parts of a discourse formed by what Michael Foucault (in *The Order of Things* and *The Archaeology of Knowledge*) terms 'discursive practice'—a process in which the thought-object of the discourse and the concepts corresponding to this object are formed by a theoretical labour upon empirically given raw material. My object in this chapter is to attempt to provide a *description* of this discourse, in order to set out its basic explanatory limitations. By viewing Parsons's texts in this way, as

components of a discourse produced by a specific process of theoretical abstraction, we can, it seems to me, most adequately achieve this objective.

13. E. Devereux, Jr., 'Parsons Sociological Theory' in M. Black, *The Social Theories of Talcott Parsons*, Prentice-Hall, 1961.
14. See G. Rocher, *Talcott Parsons and American Sociology*, Nelson, 1974.
15. T. Parsons, *The Structure of Social Action*, op. cit., 8 above, p. 730.
16. For a definition of the empiricist conception of knowledge, see L. Althusser, *Reading Capital*, New Left Books, London, 1970, p. 39.
17. *The Structure of Social Action*, op. cit., 8 above, p. 698.
18. See, for example, L. Robbins, *An Essay on the Nature and Significance of Economic Science*, London 1932; O. Lange, *Political Economy*, Vol. I, Pergamon Press, 1963, etc. For a critique of the concept, see M. Godelier, *Rationality and Irrationality in Economics*, New Left Books, London, 1972.
19. *The Structure of Social Action*, op. cit., 8 above, p. 7.
20. T. Parsons, *Towards a General Theory of Action*, Harvard University Press, Cambridge, Mass., 1951, p. 88.
21. G. Rocher, op. cit., 14 above, p. 39.
22. *Towards a General Theory of Action*, op. cit., 20 above, p. 78.
23. By 'classical' functionalism I am referring to the formulations of Malinowski, particularly in *A Scientific Theory of Culture*, Oxford University Press, London, 1960.
24. T. Parsons, 'The Pattern Variables Re-visited: A Response to R. Dubin', in *The American Sociological Review*, Vol. 25, No. 4, pp. 192–219.
25. *The Social System*, op. cit., 4 above, p. 107.
26. *The Social System*, op. cit., 4 above, p. 169.
27. 'The Pattern Variables Re-visited', op. cit., 24 above, p. 195.
28. ibid., p. 205.
29. ibid., p. 203.
30. ibid., p. 210.
31. ibid., p. 198.
32. L. Althusser, *Reading Capital*, op. cit., 16 above, p. 187.
33. C. Hempel, 'The Logic of Functional Analysis' in L. Gross, *Symposium on Sociological Theory*, Harper and Row, New York, 1959.
34. Merton's notions of 'functional alternatives', 'functional equivalents', 'functional substitutes' and those of 'manifest' and 'latent' functions are formed to 'cover' this same problem in the structural functionalist discourse. See R. K. Merton, *Social Theory and Social Structure*, Free Press, 1957.
35. T. Parsons, *Societies, Evolutionary and Comparative Perspectives*, Prentice Hall, 1966, p. 16.
36. ibid., p. 14.
37. T. Parsons, 'Some Considerations on the Theory of Social Change', *Rural Sociology*, Vol. 26, No. 3, September 1961, p. 192.
38. Parsons defines a 'strain' as follows: '*Strain* refers to a condition in the relation between two or more structured units that constitutes a tendency or pressure toward changing that relation to one incompatible with the equilibrium of the relevant point of the system'—'Some Considerations . . .' op. cit., 37 above, p. 196.

39. A. D. Smith, *The Concept of social Change*, R.K.P., London, 1973, p. 17.
40. T. Parsons, *Societies . . .*, op. cit., 35 above, p. 22.
41. T. Parsons, *Sociological Theory and Modern Society*, The Free Press, New York, 1967, essay entitled 'Evolutionary Universals in Society', pp. 490–520.
42. Particularly, N. J. Smelser, 'Mechanisms of Change and Adjustment to Change' in B. F. Hoselitz, and W. E. Moore, *Industrialisation and Society*, Mouton, The Hague, 1963.
43. B. F. Hoselitz, 'Social Structure and Economic Growth' paper written in 1953, printed as Ch. 2 in B. F. Hoselitz, *Sociological Factors in Economic Development*, Free Press, 1960.
44. See particularly, A. G. Frank, 'The Sociology of Development and Underdevelopment of Sociology' in *Latin America: Underdevelopment or Revolution*, Monthly Review Press, 1969.
45. D. McClelland, *The Achieving Society*, Princeton, Von Nostrand, 1961, p. 205.
46. D. McClelland, 'Motivational Patterns in S.E. Asia with Special Reference to the Chinese Case', *Journal of Social Issues*, Vol. 29, No. 1, January 1963.
47. Critiques of McClelland: J. H. Kunkel, 'Psychological Factors in the Analysis of Economic Development' in *Journal of Social Issues*, Vol. 29, No. 1., also 'Values and Behaviour in Economic Development' in E.D.G.C., Vol. 11, No. 4.; S. N. Eisenstadt, 'The Need for Achievement' in *Economic Development and Cultural Change*, Vol. 11, No. 4, p. 431; also, Frank, op. cit., 44 above.
48. R. Bendix, 'Tradition and Modernity Reconsidered' in *Comparative Studies in Society and History*, No. 9. See also, J. S. Gusfield, 'Tradition and Modernity' in *The American Journal of Sociology*, 1972.
49. A. G. Frank, 'The Sociology of Development', op. cit., 44 above.
50. See, particularly A. G. Frank 'The Sociology of Development . . . ' op. cit., 44 above.
51. See M. J. Levy, *Modernisation and the Structure of Societies*, Princeton University Press, 1966.

CHAPTER 2 THE SOCIOLOGY OF DEVELOPMENT—UNFOUNDED AXIOMS: THE RESTRICTED AND UNEVEN DEVELOPMENT OF THIRD WORLD ECONOMIES

1. Particularly, P. Baran, *The Political Economy of Growth*, Monthly Review Press, 1957; A. G. Frank, *Capitalism and Underdevelopment in Latin America*, Monthly Review Press, 1969 and *Latin America: Underdevelopment or Revolution*, Monthly Review Press, 1969; C. Furtado, *Development and Underdevelopment*, University of California Press, 1971; T. Dos Santos, 'The Structure of Dependence' in *American Economic Review*, May 1970.
2. In Parsons's work, see particularly, 'Evolutionary Universals in Society' in *Sociological Theory and Modern Society*, Free Press, N.Y., 1967; 'The Institutional Framework of Economic Development' *Structure and Process*

in *Modern Society*, Free Press, N.Y., 1960.

3. See A. G. Frank, *Latin America*, op. cit., 1 above, and also P. Baran, *The Political Economy of Growth*, op. cit., 1 above.
4. The term 'World Division of Labour', denotes the effects of the domination of the world by the capitalist mode of production, a domination that is economic, political and ideological. The enlarged reproductive requirements of the industrialised capitalist modes of production produce a 'polarisation' in the development of the world's productive forces between themselves and the economies of the Third World. The production relations and forces of the latter are dominated by the former, giving rise to a relatively slow development on the one hand and an 'accelerated' development of productive forces on the other. The dominance of the industrial capitalist modes of production produces a specific combination of internal relations and forces of production in the social formations of the Third World which acts to 'block' the development of their productive forces. It is this 'combination' that is our object of study in the following chapters.
5. See Frank, Baran, Furtado, Dos Santos, op. cit., 1 above.
6. Jalée has calculated that the Third World contributed only 47% of the agricultural production of what he terms the 'non-socialist' world. By this term, 'non-socialist world', Jalée is referring to all countries excluding the U.S.S.R., the countries of 'Eastern Europe', the People's Republic of China, N. Korea, N. Vietnam (the text was published in 1968), Mongolia and Cuba. See P. Jalée, *The Pillage of the Third World*, Monthly Review Press, N.Y., 1968, Chapter I, and *The Third World in World Economy*, Monthly Review Press, N.Y., 1969.
7. For further details and analyses of the extent of dependency on export crop production, see particularly, S. Amin, *Neo-Colonialism in West Africa*, Penguin African Library, 1973; P. Jalée, *The Third World in World Economy*, op. cit., 6 above; P. Bairoch, *The Economic Development of the Third World since 1900*, Methuen, 1975.
8. P. Bairoch, op. cit., 7 above, p. 43.
9. See H. Magdoff, *The Age of Imperialism*, Monthly Review Press, N.Y., 1966 and P. Jalée, *The Pillage of the Third World*, and *The Third World in World Economy*, op. cit., 6 above.
10. We will see later in the text that this whole process has, of course, occurred in a much more complex manner than is suggested here; whilst the creation of a dependent export sector required a massive restricting of the domestic sector at one stage, this restricting has the later disadvantage that any attempt to extend dependent capitalist development further into the agricultural sector runs up against the reproductive process of the 'restricted' system of production that capitalist penetration has already created.
11. P. Jalée, *The Pillage . . .*, op. cit., 6 above, pp. 19–20.
12. See the *United Nations Statistical Yearbook*, 1974, pp. 12–13.
13. B. Warren, 'Myths of Underdevelopment' in *New Left Review*, No. 81, September–October 1973.
14. Arghiri Emmanuel, 'Current Myths of Development' in *New Left Review*, No. 85, May–June 1974.
15. For example: the production of fertilisers, rubber products for the domestic

market, paper from timber extraction, chemical production tied to petroleum extraction, etc.
16. For an excellent critique of strategies for developing industries tied to raw material extraction and processing, see *Free Trade Zones and Industrialisation in Asia*, pub. by the Pacific-Asia Resources Centre, Tokyo, 1977.
17. See for example, in Table XIII, the increase in the production of 'chemical' and 'paper products'.
18. A recent survey found that as much as 30% of international trade took the form of exchanges between such subsidiaries. See Nicos Poulanzas, *Classes in Contemporary Capitalism*, New Left Books, 1975.
19. What follows relies on the analysis by J. Petras, P. McMichael, R. Rhodes, 'Industry in the Third World', *New Left Review*, No. 85, May–June 1974.
20. See Petras et al., op. cit., 19 above, pp. 99–100.
21. The problem of the specificity of the class structure that is determined in the last instance by the form of restricted and uneven development we have outlined is the object of our analysis in Chapter 13.
22. P. Bairoch, op. cit., 7 above, p. 58.
23. ibid., p. 105.
24. For example, B. Warren, in 'Myths of Underdevelopment', op. cit., 13 above.
25. P. Jalée, *The Pillage . . .* , op. cit., 6 above, p. 32.
26. The conclusions that can be drawn from such data concerning the future possible directions of uneven and restricted development are, of course, extremely limited, since the 'average' for each region conceals such broad disparities as to render them fairly meaningless. Contrast, for example, Iran with its export earnings of $6076.6 mill and its trading surplus of $17830.6 with a country such as Thailand with its equivalent earnings of $1401.6 mill and its trading deficit of $482.6 in 1973.
27. A. Emmanuel, *Unequal Exchange*, New Left Books, London, 1972.
28. C. Payer, *The Debt Trap: the I.M.F. and the Third World*, Penguin 1974; See also, T. Hayter, *Aid as Imperialism*, Penguin, 1971.

CHAPTER 3 THE SOCIOLOGY OF UNDERDEVELOPMENT: THE THEORIES OF BARAN AND FRANK

1. P. Baran, *The Political Economy of Growth*, Monthly Review Press, N.Y., 1957; *Monopoly Capital* (with Paul Sweezy) Penguin, London, 1966.
2. P. Sweezy, *The Theory of Capitalist Development*, Monthly Review Press, N.Y., 1942; *Monopoly Capital* (with P. Baran), op. cit., 1 above.
3. A. G. Frank, *Capitalism and Underdevelopment in Latin America*, Monthly Review Press, 1967; *Latin America: Underdevelopment or Revolution?*, Monthly Review Press, N.Y., 1969; *Lumpenbourgeoise: Lumpendevelopment – Dependence, Class and Politics in Latin America*, Monthly Review Press, N.Y., 1972; *Dependence and Underdevelopment: Latin America's Political Economy* (with J. Cockcroft and D. Johnson), Doubleday, N.Y., 1972.
4. For the definition of economic surplus, see P. Baran, *The Political Economy of Growth*, op. cit., 1 above, Chapter 2.

5. For the definition of economic surplus in *Monopoly Capital*, see Chapter 1 of the text, Section 2, pp. 17–26. See also Appendix of the text, pp. 355–74.
6. See P. Baran and P. Sweezy, *Monopoly Capital*, op. cit., 1 above, p. 113.
7. See Chapter 1, 'Actor and Action'.
8. K. Marx, *A Contribution to the Critique of Political Economy*, Laurence and Wishart, London, 1971, pp. 197–8.
9. A. G. Frank, *Capitalism and Underdevelopment in Latin America*, op. cit., 3 above, p. 13.
10. P. Baran, *The Political Economy of Growth*, op. cit., 1 above, p. 44.
11. ibid., p. 137 (our emphasis).
12. See K. Marx, *Capital*, Vol. 3, Progress Publishers, Moscow, 1966, pp. 790–2.
13. P. Baran, *The Political Economy of Growth*, op. cit., 1 above, p. 144.
14. ibid., p. 162.
15. ibid., p. 150.
16. This point is stated most clearly in Chapter 5 of Baran's text, *The Political Economy of Growth*, op. cit., 1 above.
17. See the texts of A. G. Frank, op. cit., 3 above.
18. A. G. Frank, *Capitalism and Underdevelopment in Latin America*, op. cit., 3 above, p. 13.
19. ibid.
20. For Frank's notion of the process of 'continuity in change', see ibid., Chapter I, pp. 12–14.
21. ibid., p. 9.
22. As far as I am aware these are the only references in Frank's texts where capitalism is defined.
23. ibid., p. 227 (our emphasis).
24. ibid., p. 7.
25. ibid., p. 227 (our emphasis).
26. ibid., p. 29.
27. ibid., p. 232.
28. See, A. G. Frank, 'Capitalist Latifundia Growth in Latin America' in *Latin America: Underdevelopment or Revolution?*, op. cit., 3 above.
29. See the recent works of P. P. Rey, E. Terray and others, particularly, P. P. Rey, *Colonialisme, néo-colonialisme et transition au capitalisme*, Maspero, Paris, 1971; E. Terray, *Marxism and 'Primitive' Societies*, translated by Mary Klopper, Monthly Review Press, New York, 1972. For an introduction to Rey's work in English, see G. Dupré and P. P. Rey, 'Reflections on the pertinence of a theory of the history of exchange' in *Economy and Society*, Vol. 2, No. 2, May 1973.
30. The problem of how to analyse these effects is examined later in the present text.
31. A. G. Frank, *Capitalism and Underdevelopment in Latin America*, op. cit., 3 above, p. 269.
32. ibid., p. 24.
33. ibid., p. 230.
34. See P. Baran, *The Political Economy of Growth*, op. cit., 1 above, p. 228.
35. For the definition of the concept of 'potential economic surplus, see ibid., p. 26. For an elaboration of its use in Baran's concept of a 'rationally

ordered society', see Chapter 8 of *The Political Economy of Growth*, op. cit., 1 above.

36. P. Baran, *The Political Economy of Growth*, op. cit., 1 above, p. 249.
37. ibid., p. 262.
38. A. G. Frank, *Capitalism and Underdevelopment in Latin America*, op. cit., 3 above, p. 29 (our emphasis).
39. ibid., p. 145.
40. For an outline of the general approach of the Sociology of Underdevelopment: see, in particular, the introduction by H. Bernstein to his text of edited readings, *Underdevelopment and Development*, Penguin, London 1973.
41. A. G. Frank, *Capitalism and Underdevelopment in Latin America*, op. cit., 3 above, p. 166.
42. ibid., pp. 56–7, 94–6.
43. On these points, see T. C. Smith, *Agrarian Origins of Modern Japan*, Stanford, 1959.
44. The simple argument that 'satellites' experience their most rapid economic growth when cut off from the metropolis is clearly refuted by the high growth rates of dependent economies mentioned in Chapter 2. The more rigorous position that national development geared towards the requirements of the domestic market is determined by the progressive removal of metropolitan dominance is, however, far from adequate, not only for explaining the case of Peronist Argentina, but also for other periods of national capitalist development, such as Nasser's Egypt, Sukarnoist Indonesia, etc. In both these cases, such phenomena as the class alliances on which political power rested and the resistance of the landowning class to capitalist penetration of agriculture or the isolation of sections of the capitalist class tied to foreign investment or imports are of primary importance, rather than a process of 'increasing contact' with the metropolis.
45. A. G. Frank, *Capitalism and Underdevelopment in Latin America*, op. cit., 3 above, p. 10.
46. ibid., p. 5 (our emphasis).
47. ibid., p. 11.

CHAPTER 4 THEORETICAL PREREQUISITES FOR AN ANALYSIS OF THIRD WORLD FORMATIONS: THESES

1. The concept of *dislocation* will be developed later in the text (notably Chapter 6). It is concerned with the effects of the transitional economic structure (the combined articulation of modes of production) on the other levels of the social formation. When one mode of production is dominant, the other levels (political, ideological) of the social formation are 'adapted' to it, in the sense that they can only operate within limits that are ultimately determined by the mode of production (in its various stages of development). The system of interventions of one level within another at any one moment in the history of a social formation is 'structured' (see Chapter 5) by these limits. In a transitional period, we have a co-existence of political and

ideological forms required for different modes of production, which are in contradiction; consequently, these forms are no longer 'adapted' to the economic structure, in the above sense. We have, therefore, 'dislocations' between the levels of the superstructure, and between these levels and the mode of production. These dislocations can only be adequately analysed by a double reference—to the structure and reproductive requirements of different modes of production.

2. For a specification of the concept of conjuncture, see L. Althusser, 'On the Materialist Dialectic' in *For Marx*, Allen Lane, London, 1969.

CHAPTER 5 SOCIAL FORMATION AND MODE OF PRODUCTION

1. See Chapter 1.
2. For an analysis of Weber's concept of totality, see P. Q. Hirst, *Social Evolution and Sociological Categories*, Allen and Unwin, London, 1976, Chapters 3–6.
3. This problem of the 'conditions of existence' of a discourse has only really been systematically approached in the work of Michael Foucault (in *The Order of Things*, Tavistock Press, London, 1970, *Madness and Civilisation*, Tavistock Press, London, 1965, *The Archaeology of Knowledge*, Tavistock Press, London, 1972, *The Birth of the Clinic*, Tavistock Press, London, 1975). Foucault states the problems of analysing the conditions of existence of the discourse very succinctly in an interview in *Cahiers Pour L'Analyse*, No. 9. (Paris, Summer 1968); selections from this interview are translated in *Theoretical Practice*, No. 3/4, Autumn, 1971. He concludes:

> What has to be brought out is the set of conditions which, at any given moment and in a determinate society, govern the appearance of statements, their preservation, the links established between them, the way they are grouped in statutory sets, the role they play, the action of values or consecrations by which they are affected, the way they are invested in practices or attitudes, the principles according to which they come into circulation, are repressed, forgotten, destroyed, or re-activated. In short, it is a matter of the discourse in the system of its institutionalisation. (*Theoretical Practice*, 3/4, p. 116).

What Foucault is attempting to do here, therefore, is to set up epistemological protocols which establish the *rules of formation* for a given set of statements—or what he terms a 'discursive formation'. He defines such formations (i) on the basis of the objects of a discourse in a particular space; thus the objects of a discourse on, for example 'madness' cannot be found in any object called 'madness' that exists through historical time, but must be sought in the social sanctions, or more particularly, legal and religious measures that *delineate* the objects of discourse on madness (see *Madness and Civilisation*, 1965, above); (ii) on the basis of the types of statements contained in a particular domain (e.g. in the clinical discourse examined in *Birth of the Clinic*); (iii) in terms of the way they use a series of concepts; (iv) in terms of the strategic possibilities offered by a particular theme (e.g. the evolutionist 'theme' in the nineteenth century).

Despite his specification of the conditions of emergence of discursive formations and his brilliant descriptions of particular discourses, there are, however, a number of serious limitations to Foucault's work—most notably the non-theorisation of transition from one discussive formation to another, the vagueness of the definition of discursive formation when this is brought to the analysis of specific raw material, the restriction of the analysis to non-scientific discourses, and the absence of any epistemological criteria for distinguishing between scientific and non-scientific discourse. On the latter points, see A. Hussein's introduction to Foucault's work in 'A Brief Resumé of the Archaelogy of Knowledge', *Theoretical Practice*, 3/4, Autumn 1971, pp. 104–7.

4. See in particular L. Althusser and Etienne Balibar, *Reading Capital*, New Left Books, London, 1970; L. Althusser, *For Marx*, Allen Lane, London, 1969.
5. K. Marx, *A Contribution to the Critique of Political Economy*, Laurence and Wishart, 1971, 'Introduction to a Critique', p. 190.
6. K. Marx, *Capital*, Vol. 2, Progress Publishers, Moscow, 1967, pp. 36–7.
7. I stress the term labour process*es*, since the dominance of particular relations of production is exercised—as I indicate below—over a combination of different divisions of labour, whose development, whilst being governed by the requirements of a specific form of extraction of surplus labour, is *uneven*, in the sense that some of them will be subsumed to a greater degree than others at any one moment in the development of a particular mode of production. It is this process of uneven subsumption that Marx is concerned with in, for example, his analysis of the transition from manufacture to machine industry in *Capital*, Vol. I, Pt. IV. This is equally the case with other non-capitalist modes of production; in the feudal mode, for example, there can exist different labour processes based variously on co-operation, labour-service, wage-labour or tenant cultivation.
8. E. Balibar, 'On the Basic Concepts of Historical Materialism' in L. Althusser et al., *Reading Capital*,op. cit., 4 above, p. 211.
9. 'Ownership' of the means of production: in Marx's work, as is well known, there is a tendency to conflate *legal ownership* of the means of production with the *ability to control and possess* the latter, which is a function determined by the structural place occupied by the *support* for this structure, 'the capitalist' or the representatives of capital. It is, of course, crucial in the analysis of any mode of production to distinguish between *possession* (the ability to put the means of production into operation which can be individual or collective), and legal ownership (the power to appropriate the means and dispose of the object of production). Both these functions of legal ownership and possession can be exercised by the property-owners themselves or by representatives (agents) acting for them. On these points, see Yves Duroux, 'Theoretical Comments' in *Sur L'Articulation des modes de production*, Cahiers de Planification, 13–14, Paris, 1970.
10. K. Marx, *Capital*, Vol. 3, Progress Publishers, Moscow, 1966, p. 791.
11. The concept 'instance', is used throughout this text for the following reason: each practice does not exist autonomously, but intervenes within other

practices, and is itself intervened in by the latter, both ultimately within limits set by the reproductive requirements of the mode of production in its dynamics. Consequently, in analysing any aspect of the structure, such as the political, we are always faced with a set of necessary interventions. Following Badiou, I have used the term 'instance' to indicate this. We can thus speak of the 'political instance' as the space within the structure in which the economic and ideological practices intervene within the limits of the political, at a particular moment in its development. See, A. Badiou, 'Le (Re) Commencement du matérialisme dialectique' in *Critique*, Paris, 1967.

12. For analyses of the lineage mode of production, see P. P. Rey, *Colonialisme, néo-colonialisme et transition du capitalisme*, Maspero, Paris, 1971; E. Terray, *Marxism and Primitive Societies*, Monthly Review Press, New York, 1972. See also, Meillassoux's text, *Anthropologie économique des Gouro de Côte d'Ivoire*, Mouton et Cie, Paris, 1964.
13. The 'elders' of the tribe: those who occupy the position of male parents and grand-parents, traced through a line of matrilineal descent in the tribal lineage.
14. This *indication* of possible ways in which the surplus labour extracted under the dominance of capitalist relations of production can be distributed is not meant in any way to be exhaustive, but is merely *illustrative*. Only a small number of the possible avenues of distribution are mentioned.
15. K. Marx, *A Contribution . . .*, op. cit., 5 above, p. 204.
16. K. Marx, *Capital*, Vol. 2, op. cit., 6 above, Ch. 5, 'The Time of Circulation', pp. 127–9; also Ch. 18, Introduction to 'The Reproduction and Circulation of the Aggregate Social Capital', p. 357.
17. K. Marx, *A Contribution. . .*, op. cit., 5 above, p. 205. For an analysis of the determination of production by processes such as circulation, that production itself determines, in the discourse of *Capital*, see R. Establet, 'Présentation du plan du Capital', in L. Althusser et al., *Lire Le Capital*, Vol. IV, Petite Collection, Maspero, Paris, 1973, pp. 47–109.
18. The notion of 'history', as a simple 'investigation of the past' (as we indicate in our discussion in this chapter), is an erroneous notion. Since the study of the past is necessarily confined to an examination of representations of what has previously existed in texts written from within contemporary ideologies, it is impossible for history to study any *real concrete historical object*—rather this must be ideologically given. Furthermore, the ideologies from within which 'history' is analysed all necessarily have their own particular notions of how to analyse historical structure and time (history as the autogenesis of an idea, a spirit; as the 'world view' of an 'age', etc.). Equally the Marxist analysis of 'history' has its own concept of historical time, which, at the economic level, is based on genealogies and dynamics of modes of production. For an outline of the *concept of historical time* used in analyses of social formations see later in this chapter.
19. See L. Althusser, 'Ideology and Ideological State Apparatuses' in Louis Althusser, *Lenin and Philosophy and Other Essays*, New Left Books, London, 1971. Also J. Rancière, 'Le Concept de Critique et la Critique de l'économie politique des "Manuscrits de 1845" au "Capital"', in L. Althusser et al., *Lire le Capital*, Vol. 3, Maspero, Paris, 1973.
20. See L. Althusser and E. Balibar, *Reading Capital*, op. cit., 4 above, pp. 91–

118, 'The errors of classical economics: Outline of a concept of historical time'.

21. By using the term 'apparatus' of the state, I am referring to those institutions through which the state carries out its functions. The state exercises its power through the government, the civil administration, the courts, the legislature, the police, etc. However, in addition to these institutions which operate primarily through the exercise of repressive means, there are a further set of apparatuses, or specialised institutions, which operate primarily through ideological means, with the objective of ensuring that the ideologies of those classes who are politically dominant become ideologies that govern individuals actions in the various areas of their social life. These 'ideological state apparatuses' can, following Althusser, be defined, as the religious, educational, familial, legal, political and trade union apparatuses.
22. By using the term class 'fraction', I am descriptively indicating that within a particular class (whose limits are defined economically, politically and ideologically), there are groups who are capable of becoming autonomous (politically or economically) from other groups in that class in particular conjunctures. For example, within the capitalist class during the period of dominance of finance capital, we can distinguish, at the economic level, commercial, banking and industrial 'fractions' which can have conflicting economic interests, and pursue disparate political objectives, despite the fact that they are constituted as a class in the overall structure of production, and will 'unify' when threatened by those classes who—as an effect of the economic structure—are necessarily in opposition to them.
23. For an analysis of this problem, see in particular, M. Fichant and M. Pecheux, *Sur l'histoire des sciences*, Maspero, Paris, 1971; D. Lecourt, *Une crise et son enjeu*, Maspero, Paris, 1973; L. Althusser, *Philosophie et philosophie spontanée des savants*, Maspero Paris, 1974; M. Castels and I. de Ipolla, 'Pratique épistomologique et sciences sociales', *Théorie et Politique*, December 1973.
24. We are referring here to the analysis of philosophical ideologies by L. Althusser in *Reading Capital*, and, of course, to Michel Foucault, whose texts, *Madness and Civilisation, Birth of the Clinic*, and *The Order of Things*, reveal very clearly the vast distance separating empirically given notions from the complex temporal processes by which the cultural formations in which they are formed were created through the inter-relation of the difference practices of the social formation. See M. Foucault, op. cit., 3 above.
25. V. I. Lenin, *Imperialism, the Highest Stage of Capitalism*, Progress Publishers, Moscow, 1968.
26. R. Hilferding, *Das Finanzkapital*, Wiener Volkabuchandlung, Wien, 1923.
27. K. Marx, *Capital*, Vol. 3, op. cit., 10 above, p. 246.
28. See the earlier section of this chapter, 'Economic practice'.
29. R. Hilferding, *Das Finanzkapital*—from a translation of part of the text given in D. K. Fieldhouse ed., *The Theory of Capitalist Imperialism*, Longman, 1969, Ch. 18, pp. 74–85.
30. K. Marx, *Capital*, Vol. 3, op. cit., 10 above, p. 252.
31. ibid., Ch. 15.
32. ibid., p. 256.

33. See, V. I. Lenin, *Imperialism, the Highest Stage of Capitalism*, op. cit., 25 above, pp. 99–102. See also V. I. Lenin, *The Nascent Trend of Imperialist Economism*, Progress Publishers, Moscow, 1969.
34. The notion of 'competitive capitalism' as a term for describing a particular stage in the dynamics of the capitalist mode of production is, of course, established at an extreme level of generality in Vol. 2 of *Capital*. Here I have been able to do no more than *indicate* that it is a stage with a particular specificity, both with regard to its form and effects. The theorisation of this 'stage' which, given the problem with the phrase 'competitive' capitalism (implying an absence of state intervention and monopolisation) I have preferred to term 'the dominance of commodity export', is an important task for the theory of capitalist penetration of non-capitalist modes of production.
35. For an excellent description of this 'period', see E. Hobsbawm, *Industry and Empire*, Penguin Books, 1969, Chapters 5–7.
36. *Capital*, Vol. 3, op. cit., 10 above, p. 237.
37. K. Marx, *Pre-Capitalist Economic Formations*, Lawrence and Wishart, London, 1964, p. 109. Also, K. Marx, *Grundrisse*, Penguin Books, 1973, p. 506.
38. For further analysis of these points, see Ch. 8. The widely accepted argument of Hobsbawm that the restriction of capitalist development in the seventeenth century was due to a prolonged economic crisis resulting from a constant lack of demand for commodities has recently been criticised by the Russian historian, Lublinskaya, who more adequately attributes the slow development of capitalist production to the difficulties experienced in transferring existing labour processes from 'hand' to 'machine' production—See A. D. Lublinskaya, *French Absolutism; The Crucial Phase*, Cambridge, 1968, Ch. 1.
39. K. Marx, *Capital*, Vol. 3, op. cit., 10 above, Parts 2 and 4.

CHAPTER 6 TRANSITIONAL SOCIAL FORMATIONS STRUCTURED BY AN ARTICULATION OF MODES OF PRODUCTION

1. The concept *articulation* is used here because, as opposed to the descriptive terms, 'co-existence' or 'combination' of modes of production, it indicates that the inter-relation of these modes has its *structural determinants*, in the reproductive requirements of co-existing capitalist and non-capitalist modes. These requirements are transformed as the capitalist mode becomes dominant, as it increasingly restricts the reproduction of the elements of the non-capitalist mode of production. Concomitantly, the articulation of modes that structures the social formation also undergoes change. This point is dealt with at length in Chapter 13.
2. See K. Marx, *Capital*, Vol. 3, Progress Publishers, Moscow, 1966, Part 8, pp. 815–84.
3. See *Capital*, Vol. 1, Part 8, on primitive accumulation; Vol. 3, Part 6, Ch. 42, 'Genesis of Capitalist Ground-Rent', Progress Publishers, Moscow, 1965 and 1966.

CHAPTER 7 THE GENEOLOGY OF THE ELEMENTS OF THE CAPITALIST MODE OF PRODUCTION WITHIN NON-CAPITALIST SOCIAL FORMATIONS

1. Most notably in the chapter of the *Grundrisse* entitled 'Forms which precede capitalist production', (pp. 471–514). See K. Marx, *Grundrisse*, Penguin Books, London, 1973. Also, *Capital*, Vol. I, Chapters 26 and 33, and Vol. 3, Ch. 31. See K. Marx, *Capital*, Vols. 1 and 3, Progress Publishers, Moscow, 1965, 1966.
2. K. Marx, *Capital*, Vol. I, Progress Publishers, Moscow, 1965, p. 714.
3. See P. P. Rey, *Les Alliances de classes*, Maspero, Paris, 1973. See also the critique of Rey on this point by A. Cutler and J. Taylor, 'Theoretical Remarks on the Transition from Feudalism to Capitalism' in *Theoretical Practice*, No. 6.
4. On these points, see W. Kula, *Theorie économique du système féodal*, Mouton, Paris, 1970. Also, A. D. Lublinskaya, *French Absolutism: the Crucial Phase 1620–9*, Cambridge University Press, 1968.
5. K. Marx, *Capital*, Vol. I, op. cit., 2 above, p. 742.
6. For an analysis of the varying forms taken by usury, see R. H. Tawney, *Religion and the Rise of Capitalism*, Penguin Books, London, 1964.
7. See the analyses of the economic historians, Hobsbawm, Hill and Hilton; notably Christopher Hill, *From Reformation to Revolution*, Penguin Books, London, 1969; also R. Hobsbawm, 'The General Crisis of the European Economy in the Seventeenth Century', *Past and Present*, Nos. 5 and 6, 1964. For a critique of this, see A. D. Lublinskaya, op. cit., 4 above.
8. On this point, see Frank's analysis on the emergence of commodity exchange between the Peruvian and Chilean economies as an effect of Spanish colonialism.
9. K. Marx, *Grundrisse*, op. cit., 1 above, p. 506.
10. Witness, for instance, the example quoted by Marx from Wakefield in Chapter 33 *Capital*, Vol. I, of the 'entrepreneur' who tried to transport his entire factory system—including his wage-labour—to the colonies, only to discover that, in the absence of any conditions compelling the workers to sell their labour-power, they left their employee to set up units based on their own labour and means of production. Despite its irony, this example raises an important issue: in colonies, such as Australia, Canada, etc., colonisation initially came up against a situation in which the preconditions for capitalist production simply did not exist, since those who emigrated could accumulate for themselves, and could remain possessors and owners of their means of production on their own land, etc. The obvious fact that one cannot simply 'export' capitalist relations of production was, initially, a serious problem for the development of capitalist production in these colonies. It was only ultimately overcome by passing laws in England that ensured that colonial lands could only be bought at a price which was sufficiently high to ensure that those who emigrated would have to work for others to accumulate sufficient wealth to buy their own land, before, in turn, employing their own wage-labour, etc. The government thus established the preconditions for developing capitalist production, by *introducing* a necessary initial separation of producers from their means of production.

CHAPTER 8 THEORISING THE NON-CAPITALIST MODE OF PRODUCTION: PROBLEMS AND PERSPECTIVES

1. Emmanuel Terray, *Marxism and 'Primitive' Societies*, Monthly Review Press, New York, 1972.
2. See ibid., p. 99.
3. See Terray's analysis of the Gouro society of the Ivory Coast in 'Historical Materialism and Segmentary Lineage-based Societies' ibid., pp. 93–184.
4. See C. Meillassoux, *Anthropologie économique des Gouro de Côte d'Ivoire*, Mouton, Paris, 1964. See also C. Meillassoux, 'From Reproduction to Production' in *Economy and Society*, Vol. I, No. 1, February 1972.
5. Centre d'Études et de Recherches Marxistes, *Sur le 'mode de production asiatique'*, Éditions Sociales, Paris, 1969. *Sur les sociétés pré capitalistes*, Éditions Sociales, Paris, 1970.
6. See, for example, the articles in the C.E.R.M. text on the Asiatic mode (op. cit., 5 above), which show that this mode has been dominant at one historical period or another in every continent except N. America.
7. K. Marx, *Capital*, Vol. 3, Progress Publishers, Moscow, 1966, p. 791.
8. These texts are collected in the volume, K. Marx and F. Engels, *On Colonialism*, Progress Publishers, Moscow, 1968.
9. On the relation of *Capital* to the *Grundrisse* and of both to Marx's original schema for his theoretical work on the capitalist mode of production, see R. Rosdolsky, *Zür Entstehungsgeschichte des Marxischen "Kapital"* (two volumes), Europäische Verlagsanstalt, Wien, Frankfurt-am-Main, 1968.
10. K. Marx, *Grundrisse*, Penguin Books, London, 1973, p. 486.
11. See, for example, analyses of non-capitalist modes of production in the journal, *Critique of Anthropology*, London.
12. For the work of Godelier on non-capitalist formations, see, particularly, *Horizon, trajets Marxistes en anthropologie*, Maspero, Paris, 1973.
13. M. Godelier, 'System, Structure and Contradiction in Capital', *Socialist Register*, London, 1967, pp. 91–114.
14. ibid., p. 100.
15. ibid., p. 105.
16. ibid., p. 106.
17. ibid., p. 106.
18. G. Sofri, *Il mode di produzioni asiatico*, Einaudi, Turino, 1963, p. 72.
19. For a critique of Godelier on this particular point, see A. Hussain, 'Godelier's Rationality and Irrationality in Economics', in *Theoretical Practice*, London, No. 7/8, pp. 86–94.
20. See M. Godelier, 'Preface' to C.E.R.M. *Sur les sociétés précapitalistes*, op. cit., 5 above, pp. 13–142.
21. See M. Godelier, 'La notion de 'mode de production asiatique' et les schemas Marxistes d'évolution des sociétés', in C.E.R.M. *Sur la 'mode de production asiatique'*, op. cit., 5 above, pp. 47–100.
22. These two quotes are from Godelier's analysis of the Asiatic mode of production, ibid., pp. 50 and 85 respectively.
23. M. Godelier, 'Anthropologie et économie', in M. Godelier, *Horizon trajets marxistes en anthropologie*, Maspero, Paris, 1963.
24. ibid.

25. B. Hindess and P. Q. Hirst, *Pre-capitalist Modes of Production*, Routledge, London, 1975.
26. B. Hindess and P.Q. Hirst, *Mode of Production and Social Formation*, Macmillan, London, 1977.
27. ibid., p. 22.
28. See John G. Taylor, 'Pre-capitalist Modes of Production' in *Critique of Anthropology*, No. 4/5 Autumn 1975, and No. 6, Spring 1976.
29. B. Hindess and P. Q. Hirst, *Pre-capitalist Modes* . . . , op. cit., 25 above, p. 10.
30. ibid., p. 193.
31. See Chapter 5.
32. For an outline of this mode of production, see Chapter 5.
33. See B. Hindess and P. Q. Hirst, *Pre-capitalist Modes* . . . , op. cit., 25 above, pp. 21–78.
34. ibid., p. 44.
35. ibid., p. 45.
36. ibid., p. 53.
37. ibid., p. 48.
38. ibid., p. 65.
39. ibid., p. 273.
40. ibid., p. 274.
41. See Chapter 7.
42. See B. Hindness and P. Q. Hirst, *Pre-capitalist Modes* . . . , op. cit., 25 above, p. 274.
43. L. Althusser and E. Balibar, *Reading Capital*, New Left Books, London, 1970, pp. 292–3.
44. See Chapter 13.

CHAPTER 9 CONCEPTUALISING NON-CAPITALIST MODES OF PRODUCTION: THE ASIATIC MODE

1. This assumption has, of course, been severely criticised by many theorists who have concluded that the social formations pre-existing capitalist penetration of China and India cannot be adequately analysed with Marx's concept of the Asiatic mode. On this point, see C.E.R.M. *Sur le 'mode de production asiatique'*, Éditions Sociales, Paris, 1969. Indeed, as I indicate below, there are only a few social formations in S.E. Asia that correspond to Marx's description of formations dominated by the Asiatic mode, and it is these that I have taken as 'raw material' for the theorisation of this mode's structure and reproduction. For Marx's writings on India and China, see K. Marx and F. Engels, *On Colonialism*, Progress Publishers, Moscow, 1968.
2. See, K. Marx, *Grundrisse*, Penguin Books, London, 1973, pp. 471–574.
3. C.E.R.M. op. cit., 1 above.
4. See the critique of Godelier's texts on non-capitalist social formations in Chapter 8.
5. See, in particular, Godelier's articles, 'Le Concept de formation économique et social; l'example des Incas', in *Horizon, trajets Marxistes en*

anthropologie, Maspero, Paris, 1973, pp. 83–92, and 'La notion de 'mode de production asiatique' et les schemas Marxistes d'évolution des sociétés', in C.E.R.M., op. cit., 1 above, pp. 47–100.

6. By using the term 'empiricist model' here, I am indicating that these notions of the Asiatic mode are confined within an empiricist conception of knowledge, in which a model is conceived as a reconstruction of the ordering of facts observed and described in reality. This reconstruction is presumed to highlight the importance of some of these facts for the purpose of explaining the existence of the phenomenon. The validity of the model is then determined by its 'fitting' the facts when the phenomenon it purports to explain is encountered again 'in reality'. For a critique of the empiricist conception of knowledge, see Ch. 1 of this text. For a critique of this notion of model, see A. Badiou, *Le Concept de modèle*, Maspero, Paris, 1970, and B. Hindess, 'Materialist Mathematics' *Theoretical Practice*, 3/4, Autumn, 1971.
7. K. Wittfogel, *Oriental Despotism*, Yale University Press, 1957.
8. E. Leach, 'Hydraulic Society in Ceylon', *Past and Present*, No. 15.
9. K. Polanyi, *Dahomey and the Slave Trade*, A.E.S. Monograph, No. 42, University of Washington Press, 1966.
10. For the Angkorian Kingdoms, see L. Sedov, 'La Société Angkorianne et le problème du mode de production asiatique', in C.E.R.M., op. cit., 1 above, pp. 327–43; E. T. Aymonier, *Histoire de l'ancien Cambodge*, Paris, 1920, and *Le Cambodge*, Paris, 1900–4; A. Dauphin-Meunier, *Histoire de Cambodge*, Paris, 1961; B. P. Groslier, *Angkor et la Cambodge au XVIe siècle d'après les sources portugaises et espagnoles*, Paris, 1968; A. Migot, *Les Khmers, des origines d'Angkor au Cambodge d'aujourd'hui*, Paris, 1960; C. B. Walker, *Angkor Empire*, Calcutta, 1955; L. P. Briggs, *The Ancient Khmer Empire*, Philadelphia, 1951.
11. *On the Indonesian Empires*, see: T. S. Raffles, *History of Java*, London, 1817; G. Coèdes, *The Indianised States of S.E. Asia*, Honolulu, 1968; R. von Heine-Geldern, *Conceptions of State and Kingship in Southeast Asia*; Ithaca, N. York, 1963; T. G. Pigeaud, *Java in the Fourteenth Century: A study in Cultural History* (5 volumes), The Hague, 1960–3; F. M. Schnitger, *Forgotten Kingdoms in Sumatra*, E. J. Brill, Leiden, 1939; O. W. Wolters, *Early Indonesian Commerce: A Study of the Origins of Srivijava*, Ithaca, New York, 1967.
12. On Vietnam during this period, see: Nguyen Thanh-Nha, *Tableau économique du Vietnam aux XVIIe et XVIIIe siècles*, Éditions Cujas, Paris, 1970; Jean Chesnaux, *Contribution à l'histoire de la nation Vietnamienne*, Paris, Éditions Sociales, 1955; Le Thanh Khoi, *Le Vietnam: Histoire et civilisation*, Paris, Editions de Minuit, 1955; Joseph Buttinger, *The Smaller Dragon, Praeger, New York, 1958*; Nguyen Van Huyen, *La Civilisation annamitée*, Hanoi, 1944; J. Riche, 'La Cochinchine au XVIIIe siècle', *Revue d'Europe et des Colonies*, Paris 1906; Vu Quoc Thue, *L'économie communaliste du Viet-Nam*, ed. De Droit, Paris, 1951; L. Cadière, *Résumé de L'histoire d'Annam*, Quihnon, 1911; G. Gosselin, *L'Empire D'Annam*, Paris, 1957; C. B. Maybon, *Histoire Moderne du Pays d'Annam (1592–1820)*, Paris 1920; C. E. Rouger, *Histoire Militaire et Politique et l'Annam et du Tonkin depuis 1799*, Paris, 1906.

13. For these descriptions, see K. Marx and F. Engels, *On Colonialism*, op. cit., 1 above, pp. 35–41 and 81–7.
14. A. V. Chayanov, *On the Theory of Peasant Economy*, ed. D. Thorner, Irwin, Homewood Ill., 1966.
15. ibid., p. 6.
16. On this point, see Nguyen Thanh-Nha, op. cit., 12 above, Pt. 1, Chapter 2 and Pt. 2, Chapter 1, Section 2.
17. On this point see, particularly, Leonid Sedov, 'La société angkorienne et la problème du mode de production asiatique' in C.E.R.M., op. cit., 1 above, and Le Thanh-Khoi and Nguyen Thanh-Nha, op. cit., 12 above.
18. On this point, see Nguyen Thanh-Nha, op. cit., 12 above, Pt. 1, Section 3, pp. 59–73.
19. Ion Banu, 'La formation sociale 'asiatique' dans la perspective de la philosophie orientale antique' in C.E.R.M. op. cit., 1 above; see Nguyen Thanh-Nha, op. cit., 12 above; L. Thanh-Khoi, op. cit., 12 above.
20. See Nguyen Thanh-Nha, op. cit., 12 above, Pt. 1, Section 3, pp. 59–73.
21. For a succinct account of such an ideological definition and its effects, see Banu, op. cit., 19 above.
22. On this point, see Nguyen Thanh-Nha, op. cit., 12 above.
23. For excllent accounts of this ideological conception of kingship, see, particularly, G. Coedes, *The Indianised States of S.E. Asia*, op. cit., 11 above, and R. von Heine-Geldern, *Conceptions of State and Kingship in Southeast Asia*, op, cit., 11 above.
24. K. Marx and F. Engels, *On Colonialism*, op. cit., 1 above, p. 37.
25. K. Marx, *Grundrisse*, op. cit., 2 above, p. 473.
26. ibid., pp. 473–4.
27. Such a claim is made notably by G. Lichtheim in his article, 'Marx and the Asiatic Mode of Production', in *St. Anthony's Papers*, No. 14, 1963.
28. For this line of critique, see, particularly, D. Fernbach in his introduction to *Surveys from Exile*, Marx's political writings, Vol. 2, Penguin Books, 1973, p. 24.

CHAPTER 10 THE EFFECTS OF CAPITALIST PENETRATION ON NON-CAPITALIST MODES OF PRODUCTION: PENETRATION UNDER THE DOMINANCE OF MERCHANTS' CAPITAL

1. For the function of merchants' capital in the genealogy of the capitalist mode of production, in the particular case of the transition from feudalism to capitalism in Europe, see Chapter 7.
2. See Chapter 5.
3. See F. Engels, letter to Bebel, 11 December 1884, in K. Marx and F. Engels, *Selected Correspondence*, Progress Publishers, Moscow, 1965, pp. 379–82.
4. C. Furtado, *Economic Development of Latin America*, Cambridge University Press, 1970, pp. 14–15.
5. For an analysis of the 'non-economic' effects of plantations as units of production, see Lloyd Best, 'Outline of a model of a pure plantation economy', *Social and Economic Studies*, September 1968. Also, E. T.

Thompson, 'The Plantation as a Social System', in *Plantation Systems of the New World*, ed. Pan American Union. Also, G. Beckford, *Persistent Poverty*, Oxford University Press, London, 1972.

6. For a description of this 'ideological blending', see, for example, A. Guerrero, *Philippine Society and Revolution*, Ta Kung Pao, Hong Kong, 1971.
7. For analyses of the effects of Dutch colonialism on the agrarian sector of the Indonesian economy, see: J. S. Furnivall, *Netherlands India: A Study of Plural Economy*, Macmillan, New York, 1944. Also, J. S. Furnivall, *Colonial Policy and Practice: A Comparative Study of Burma and Netherlands India*, Cambridge, 1948; and C. Geertz, *Agricultural Involution*, University of California Press, 1963.
8. For analyses of the contemporary situation in Java's agricultural sector, with reference to the continuing reproduction of the economic effects of colonialism, see C. Geertz, op. cit., 8 above. Also, Rex Mortimer, 'Indonesia: Growth or Development?' in *Showcase State: the Illusion of Indonesia's 'accelerated modernisation'*, ed. Rex Mortimer, Angus and Robertson, London, 1973, pp. 61–6.
9. The reasons for this restricting of the extension of capitalist production during the seventeenth and up to the mid-eighteenth centuries have been the subject of a recent debate between those—such as Hobsbawm—who argue that the 'general economic crisis' in seventeenth-century Europe was produced largely by the limited domestic market and by the inability to ensure a growing demand for capitalist commodities, and those—such as Lublinskaya—who attribute the slow development of capitalist production to the difficulties it experienced in technically transforming existing production in the labour processes. As Lublinskaya states: 'The technical and economic progress, described by Marx, which took place within the manufactories themselves, was not only a complex process, but also a prolonged one, for the changes accomplished in the sphere of the making of instruments of labour necessarily passed—being for a long time limited by the general slowness of development which was characteristic of manufacture with its hand labour—through a series of successive stages from production completely by hand to production completely by machinery.' (Lublinskaya, *French Absolutism . . .*, p. 75). It seems to me—for reasons which I am unable to go into here—that Lublinskaya's critique of Hobsbawm's analysis clearly points to a number of limitations in it, limitations which are not to be found in Lublinskaya's theorisation of this period. See E. Hobsbawm, 'The General crisis of the European Economy in the Seventeenth Century', *Past and Present*, Nos. 5 and 6, 1954. For Lublinskaya's critique, see A. D. Lublinskaya, *French Absolutism: The Crucial Phase 1620–1629*, Cambridge University Press, 1968, Ch. 1, pp. 38–75.
10. On this point see, A. D. Lublinskaya, op. cit., 10 above, Ch. 1, pp. 4–81.
11. Craft guilds were able to restrict the development of capitalist production during this period by restricting the number of workers that could be employed in any one unit, by their receiving preferential treatment for their raw material supply, and by their being supported by European governments.

12. Despite the fact that sections of the peasantry were, in many areas, thrown off their land, there were (for the reasons given above) few units of production which they could enter as wage-labour. Consequently, armies of unemployed drifted aimlessly around the countryside—an effect of the 'blocked' transition characteristic of the Spanish social formation during this period.
13. See E. Hobsbawm, op. cit., 10 above, on this point.
14. For an analysis of the economic effects particular to imperialist penetration, see Chapter 12.
15. Here we should note that throughout this section we have been concerned solely with the economic forms set up by penetration under the dominance of merchants' capital, and not with the so-called problem of 'plantations' in general. Many authors mistakenly combine these very different forms set up at different stages of capitalist penetration under the same analytical rubric—'the plantation'. They thereby conflate the effects of mercantile penetration with the systems of production established during imperialist penetration. The two types are, of course, very different. The latter is characterised by a radical separation of direct producers from their means of production, by means of an enforced wage-labour in which the peasantry are thrown off their land to work exclusively on a *capitalist* agricultural enterprise. Conversely, the former—as we have seen—*reinforces* the existing unity of direct producers with their means of production, setting up patterns of land-ownership and tenure that become barriers to the emergence of wage-labour, which are profoundly 'anti-capitalist'.

    The term 'plantation' seems, therefore, to act as something of a residual category, into which all agricultural systems that require a labour-intensive exploitation of labour on an extensive scale are placed, despite the profound differences that exist between specific forms of production developed by very different stages of penetration.
16. On this point, see Chapter 5.
17. Although we have only analysed the effects of mercantile capitalist penetration on Latin American and S.E. Asian formations, and not in relation to African formations, it is clear that similar conclusions, on the effect of reinforcing existing non-capitalist relations can be reached, in the case, for example, of Portuguese intervention in West-Central Africa. On this point, see G. Dupré and P. P. Rey, 'Reflections on the pertinence of a theory of the history of exchange' in *Economy and Society*, Vol. 2, No. 2, May 1973. Also, E. Terray, 'Long-distance exchange and the formation of the State' in *Economy and Society*, Vol. 3, No. 3, August 1974. See also P. P. Rey, *Colonialisme, néo-colonialisme et transition de capitalisme*, Maspero, Paris, 1971, Part II, Chapters 3 and 4.

CHAPTER 11 THE EFFECTS OF PENETRATION UNDER THE DOMINANCE OF COMMODITY EXPORT

1. See Eric Hobsbawm, *Industry and Empire*, Penguin Books, 1969, particularly Chapters 1–7. See also, D. S. Landes, *The Unbound Prometheus: Technological Change and Industrial Development in Western Europe from*

1750 to the present, Cambridge University Press, 1970, Chapters 2 and 3.
2. On merchant-artisan 'putting-out' industries in the transition to capitalism, see Chapters 5 and 10.
3. See Chapter 5.
4. See Chapter 5.
5. For excellent accounts of the effects of British colonial rule on the Indian economy, see Marx's articles: 'The British Rule in India', 'India', and 'The Future Results of the British Rule in India', in *New York Daily Tribune* 25 June 1853, 5 August 1853 and 22 July 1853. These articles are contained in the collection *On Colonialism* by K. Marx and F. Engels, Progress Publishers, Moscow, 1968.
6. For an outline of this mode of production, see Chapter 5.

CHAPTER 12 IMPERIALISM AND THE SEPARATION OF DIRECT PRODUCERS FROM THEIR MEANS OF PRODUCTION

1. See Chapter 5.
2. For example: the extent of the dominance of the political representatives of the banking and industrial capitalist fractions in the state apparatus, the emergence and pervasiveness of national chauvinist and racist ideologies, etc.
3. See Chapter 7.
4. K. Marx, *Capital*, Vol. 3, Progress Publishers, Moscow, 1966, p. 334.
5. For detailed data on the increases in the railway networks in colonised Africa and Asia during this period, from which these statistics are taken, see R. Luxemberg, *The Accumulation of Capital*, Routledge, London, 1971, p. 420.
6. See Chapter 13.

CHAPTER 13 THE EMERGENCE OF AN ARTICULATION OF MODES OF PRODUCTION AND ITS EFFECTS ON THE STRUCTURE OF THE THIRD WORLD FORMATION

1. For definitions of the concepts formal and real subsumption of the productive forces in a given mode of production, see Chapter 5.
2. For a definition of the concept of social formation, see Chapter 5.
3. For an analysis of these processes in one particular period of industrial capitalist crisis, see J. G. Taylor, *The Indonesian Economy during the 1930's Depression*, M. A. Thesis, School of Oriental and African Studies, University of London, September 1970. See also, A. G. Frank, 'Capitalist Development of Underdevelopment in Chile' in *Capitalism and Underdevelopment in Latin America*, Monthly Review Press, New York, 1967.

4. One of the clearest examples of this strategy for promoting economic growth is that of Brazil, where the state has actively pursued policies of redistributing income from poor to rich in order to extend the internal market for domestically produced department 11a goods. See J. Serra, 'The Brazilian "Economic Miracle"' in J. Petras, ed., *Latin America: From Dependence to Revolution*, New York, 1973.
5. See Geoffrey Kay, *Development and Underdevelopment: A Marxist Analysis*, Macmillan, London, 1975, p. 153.
6. For analyses of the limitations of import-substitution, see N. H. Leff and A. D. Netto, 'Import substitution, foreign investment and international disequilibrium in Brazil' in *The Journal of Development Studies*, Vol. 2, April 1966; also C. Furtado, 'Industrialisation and Inflation', *International Economic Papers*, No. 12, 1967; also J. L. Lacroisse, 'The Concept of Import Substitution in the theory of economic development' in *Cahiers économiques et sociales*, June 1965. See also, R. B. Sutcliffe, *Industry and Underdevelopment*, Addison Wesley, 1971, p. 268.
7. See Chapter 2, p. 56.
8. For a critique of the presumed benefits of co-ownership in the industrial sector of Third World economies, with particular reference to S.E. Asia, see *Free Trade Zones and Industrialisation in Asia*, pub. by the Pacific-Asia Resources Centre, Tokyo, 1977.
9. For a description of the dependence of the peasantry upon the agricultural estates in Latin American formations, see Chapter 10.
10. See, for example, the case of the Chilean social formation during the years 1964–70, analysed later in this chapter.
11. This is evidenced in the massive increases in the manufacturing output of Third World economies, as indicated in the data given in Chapter 2.
12. For examples of strategies combining large amounts of foreign investment with low wages to produce high growth rates beginning in the mid-1960s see S. Korea (growth rate from 1964–8, 65.1 %), Taiwan (growth rate from 1964–8, 27.2 %). Data from U.N.C.T.A.D., *Trade in Manufactures of Developing Countries*, 1969.
13. On these points, see A. G. Frank, 'Capitalist Development of Underdevelopment in Chile', op. cit., 3 above.
14. For the concept of the trader as a 'linkman' or 'broker', between differing systems of production, see Norman Long, 'Structural dependency, Modes of production and economic brokerage in rural Peru', in Ivor Oxaal et al., *Beyond the Sociology of Development*, Routledge and Kegan Paul, London, 1975, pp. 253–83.
15. See Chapter 5, pp. 122–3.
16. See Chapter 12.
17. For an outline of the lineage mode, see Chapter 5. For analyses of the lineage mode in particular West-Central African formations, see especially the works of Meillassoux and Terray on the Ivory Coast, and of Rey on Congo-Brazzaville, cited elsewhere in this text, chapters 5, 8 and 12.
18. On this point, see Chapter 12, pp. 209–11.
19. For an account of this process in some detail, see P. P. Rey, *Colonialisme, néo-colonialisme et transition au capitalisme*, Maspero, Paris, 1971, Section 3, Chapter III.

20. For a short definition of absolute and differential ground rent, and the specificity of the latter to the capitalist mode of production, see A. J. Cutler and J. G. Taylor, 'Theoretical Remarks on the transition from feudalism to capitalism', in *Theoretical Practice*, No. 6, 1971.
21. The size of the service sector in Third World economies can be gauged by the fact that employment in this sector for Third World economies as a whole increased from 5.5 % in 1920, to 6.1 % in 1930, to 8.9 % in 1950, and 9.6 % in 1960; in 1970, combined employment in trade, transport, communications, banking and services totalled 21.0 %. See P. Bairoch, *The Economic Development of the Third World since 1900*, Methuen, London, 1975. Recent figures for service sector employment in particular countries are even more revealing: Zambia (1969) 43 %; Venezuela, (1971) 53 %; S. Korea (1971) 34 %; Indonesia (1971) .24 %. See P. Bairoch, pp. 246–9.
22. For example, in the distributive and retailing sectors of the urban areas.
23. For analyses of the conditions of existence of the urban semi-proletariat in Third World formations, see the two texts by Samir Amin, *Accumulation on a World Scale*, Monthly Review Press, New York, 1974, and (more particularly), the essays in *Unequal Development*, published by the Harvester Press, U.K. 1977. For descriptions of the living conditions of the semi-proletariat in particular Third World formations, see, for example, Charles Bettelheim, *India Independent*, Monthly Review Press, New York, 1971, Chapters 5, 12. See also, Mahmoud Hussein, *Class Conflict in Egypt, 1945–70*, Monthly Review Press, New York, 1973, Chapter I.
24. For example: in the formations of Egypt, India and Indonesia, the income groups within the working class in the major cities in temporary employment form over one-half of the total urban population.
25. See also Chapter 2, for a more detailed exposition of 'restricted and uneven development'.
26. For example, Iran, the Phillipines, India.
27. By using the term, 'oppressive apparatus' of the colonial state, I am referring to those institutions—the government, the administration, the army, the police, the courts, etc.—which, as distinct from the ideological apparatuses of the state, function in the last instance by suppression.
28. See, for example, Samir Amin, *Unequal Development*, op. cit., 23 above, and, G. Arrighi, 'International Corporations, Labour Aristocracies, and Economic Development in Tropical Africa' in R. I. Rhodes, ed., *Imperialism and Underdevelopment: A Reader*, Monthly Review Press, New York, 1970.
29. For an analysis of the ideological framework in which these revolts occurred, see, for example, Sartono Kartodirdjo, 'Agrarian Radicalism in Java—its Setting and Development' in Claire Holt, ed., *Culture and Politics in Indonesia*, Ithaca, New York, 1972. See, also Sartono Kartodirdjo, *Protest Movements in Rural Java*, Oxford University Press, 1973.
30. On this amalgamation, as represented in the ideas of the Indonesian Communist Party (PKI) and in those of the various political groupings that eventually became the Indonesian Nationalist Party (PNI), see, R. McVey, *The Rise of Indonesian Communism*, Cornell University Press, Ithaca, New York, 1968. Also, R. McVey, in her introduction to Sukarno's text,

Nationalism, Islam and Marxism, Cornell University Press, Ithaca, 1969. Also, F. Cayrac-Blanchard, *Le Parti Communiste Indonésien*, Armand-Colin, Paris 1973, pp. 25–32. On the effect of this amalgamation of two different ideological forms on the post-war development of the PKI, see Rex Mortimer, *Indonesian Communism under Sukarno*, Cornell University Press, Ithaca, 1974.

31. Notably Samir Amin, in *Unequal Development*, op. cit., 23 above.
32. For Marx's descriptions of the 'lumpenproletariat' and its political role in nineteenth-century French capitalism, see K. Marx, 'Class Struggles in France' and 'The Eighteenth Brumaire' in K. Marx and F. Engels, *Selected Works* (2 volumes), Progress Publishers, Moscow, 1962, Vol. 1, pp. 118–242 and pp. 234–344 respectively.
33. On the concept of Bonapartism, see N. Poulantzas, *Political Power and Social Classes*, New Left Books, London, 1973, particularly, Pt. 4, Chapter 3.
34. For an analysis of this 'political ambivalence' in the case of one particular social formation, see M. Hussein, op. cit., 23 above.
35. On this extension of capitalist production, see Chapter 2, pp. 53–60.
36. For an excellent critique of this naïve position, see Fei-Ling, *Proletarian Culture in China*, published by the Association for Radical East Asian Studies, London, 1972, Chapter 2.
37. See Chapter 10, pp. 53–60.
38. See later in this chapter.
39. For a good example of a text that somewhat dogmatically applies conclusions derived from analyses of the differentiation of the peasantry in social formations of a very different character, see A. Guerrero, *Philippine Society and Revolution*, Ta Kung Pao Publishers, Hong Kong, 1971.
40. C. Baudelot, R. Establet, J. Malemort, *La petite bourgeoisie en France*, Maspero, Paris, 1974.
41. On the role of merchants' capital in linking different systems of production, see A. G. Frank, 'Capitalist Development of Underdevelopment in Chile', op. cit., 3 above, See also, 'Capitalist Development of Underdevelopment in Brazil' in the same volume. For a further analysis of this role of merchants' capital, see Norman Long, 'Structural dependency . . .', op. cit., 14 above.
42. Perhaps the clearest example of opposition by a large petty-bourgeoisie to a programme of nationalisation and capitalist land reform in a Third World formation being a major contributory factor to the failure of these policies is that of Chile under Allende, from 1970–3.
43. For a critique of the assumption of the presence and 'progressive' role of the national capitalist class in Latin American formations, see A. G. Frank, *Latin America: Underdevelopment or Revolution?*, Monthly Review Press, New York, 1969, Section IV, particularly Chapters 14 and 23.
44. See earlier in this chapter.
45. On this point, see J. Petras, *Politics and Social Forces in Chilean Development*, University of Californian Press, 1969, Chapters 1–4.
46. This inflation of the administrative apparatus of the state during the post-war period is indicated by the following figures, for particular countries.

The Increase in Wage-earning and Salaried Clerical Workers For Selected Countries, 1967–1974*

Countries	*1964*	*1974*
Egypt	244,120	410,388
Iran	62,841	191,591
Ghana	42,220	81,765
Philippines	278,000	497,000
Venezuela	177,075	245,653

Source: *U.N. Yearbook of Labour Statistics*, 1964, 1974.

* It is extremely difficult to obtain these statistics from U.N. data with any consistency, since many years are omitted for many countries. Selection is, therefore, necessarily restricted.

47. The extension of the educational system in Third World Formations in general during the period following the Second World War is indicated by the following data:

1. *Level of illiteracy amongst the population of 15 years and above in Third World countries*

1900	–	± 80 %
1930	–	± 76 %
1950	–	± 74 %
1960	–	65 %
1970	–	56 %

2. *Changes in numbers of students at secondary and higher level of education (000), for Third World countries*

(a) *General, Vocational and Teacher Training*

1950	–	7,600
1960	–	182,000
1970	–	442,500

(b) *Universities and Other Institutions of Higher Education*

1950	–	940
1960	–	2100
1970	–	5600

(*Source*: P. Bairoch, *Economic Development* . . ., op. cit., 21 above, pp. 137 and 141).

48. On this point, see P. P. Rey, *Colonialism* . . ., *op. cit.*, 19 above, Section 4, Chapter 7.
49. The plantation sector only developed after 1870, with tobacco plantations in N. Sumatra and rubber plantations in S. Sumatra. In Java, export crops continued to be produced on small peasant plots, and collected for the Dutch companies by the village headmen, although plantations cultivating

coffee, tea and cinchona were established in the more elevated regions in the centre of the island during the period 1870–1900. Plantations existed in other areas of Java, but their land area was relatively small compared with the amount of land devoted to small peasant holdings.

50. For an analysis of the economic and political dominance of the landowning class in Indonesia during the post-independence period, see E. Utrecht, 'Land Reform and Bimas in Indonesia', in the *Journal of Contemporary Asia*, Vol. 3, No. 2, 1973, pp. 149–64. See also, E. Utrecht, 'Land Reform in Indonesia', in *Bulletin of Indonesian Economic Studies*, November, 1969, pp. 71–8; F. Cayrac-Blanchard, *Le Parti communiste Indonésien*, Armand Colin, Paris, 1973, Chapter 2; R. W. Franke, 'Hunger for Profit: Ten Years of Food Production Failure' in M. Caldwell, ed., *Ten Years Military Terror in Indonesia*, Spokesman Books, 1975; L. H. Palmier, *Social Status and Power in Java*, London, Athlone Press, 1970.
51. For an account of this comprador class, and its inter-relation with the Indonesian military, see R. Mortimer, 'Indonesia—Growth or Development' in *Showcase State: the Illusion of Indonesia's 'accelerated modernisation'*, ed. Rex Mortimer, Angus and Robertson, London, 1973, pp. 51–66.
52. For a description of this process during the mid and late fifties, see, Clifford Geertz, *Peddlers and Princes*, University of Chicago Press, London, 1963.
53. For an analysis of the trading class in Java, and the importance of the Chinese population within it, see D. E. Willmott, *The Chinese of Samarang*, Ithaca Press, New York, 1961; see also, V. Purcell, *The Chinese in Southeast Asia*, London, 1965; G. McT. Kahin, *Nationalism and Revolution in Indonesia*, Ithaca, 1952.
54. In Indonesia, during the colonial period, the urban proletariat never formed more than 3% of the total population.
55. This was evidenced in the increasing migration from rural to urban areas during the twentieth century, and in the high levels of urban unemployment that have characterised the Indonesian economy since independence in 1949. For an analysis of seasonal employment and temporary employment in urban areas, see B. Higgins, *Indonesia's Economic Stabilisation and Development*, New York, Institute of Pacific Relations, 1957. Also, D. Paauw, *Financing Economic Development: the Indonesian Case*, the Free Press, Glencoe, 1960. It is estimated that present levels of unemployment in Indonesia run between 12 and 15 million. On this latter point, see Jean-Claude Pomonti, 'Indonesia's Army Guided Democracy, II: Unemployment in an expanding economy' in *Le Monde Weekly*, 22–28 July 1971.
56. For an excellent description of the entrepreneurial ideologies of these classes, and their inter-relations with Indonesian Islam, see C. Geertz, *The Religion of Java*, New York, the Free Press, 1960.
57. For analyses of the syncretic Islamic-animist 'abangan' framework of values governing the Javanese peasantry's daily lives, see, C. Geertz, op. cit., 52 above. Also Claire Holt, ed., *Culture and Politics . . .*, op. cit., 29 above, notably the paper by Benedict O'Gorman Anderson, 'The Idea of Power in Javanese Culture'.
58. On the political history of the 'Guided Democracy' Period, see D. Lev, *The Transition to Guided Democracy*, Cornell University Modern Indonesia

Project, 1966. Also J. Pluvier, *Confrontations*, Oxford University Press, 1965.

59. For a description of these proposals, see E. Utrecht, 'Land Reform in Indonesia', op. cit., 50 above.
60. The opposition from the comprador class increasingly took a military form, as army men increasingly came to occupy comprador roles in the economic strcuture. On this point, see R. Mortimer, 'Indonesia – Growth or Development', op. cit., 51 above, and *Indonesian Communism under Sukarno*, Ithaca Press, 1974, Chapter 6.
61. For an analysis of Sukarnoism as a populist form, see R. Mortimer, *Indonesian Communism* . . . , op. cit., 60 above, Chapters 2 and 9. See also R. McVey, Introduction to *Nationalism* . . . , op. cit., 30 above.
62. For an account of the nationalisation measures proposed and implemented during the Guided Democracy period, see J. G. Taylor, 'The Economic Strategy of the New Order' in *Repression and Exploitation in Indonesia* Spokesman Books, 1974, pp. 15–17.
63. The events of the coup have been subject to much detailed empirical analysis, but perhaps the most adequate account still remains that elaborated by B. O'G. Anderson and R. McVey in their 'Preliminary Analysis of the October 1st Coup in Indonesia', published by the Cornell Modern Indonesian Project, New York, 1971.
64. For texts on this point, see J. Petras, *Politics and Social Forces in Chilean Development*, University of California Press, 1969, Chapters 1–4. I. Roxborough, P. O'Brien and J. Roddick, *Chile: The State and Revolution*, Macmillan, London, 1977, Chapters 1–3; also J. Petras, 'Christian Democracy in Chile', *New Left Review*, No. 54, March/April 1969; R. R. Kaufman, *The Politics of Land Reform in Chile*, Harvard University Press, Cambridge, Mass., 1971.
65. This period of industrialisation is analysed succinctly in J. Petras, *Politics and Social Forces in Chilean Development*, op. cit., 64 above, Chapters 1–2. The statistical data that follow are taken from Petras's text.
66. See ibid., p. 43.
67. See ibid., p. 55, Table 22.
68. ibid., p. 64.
69. According to government statistics, some 50% of Chilean workers were employed in firms with a labour force of less than 5 employees during the Frei period.
70. Chile's foreign debt (both public and private) rose from $569 to $1869 million between 1958 and 1964. On this point, see I. Roxborough et al., *Chile: the State and Revolution*, op. cit., 64 above, p. 38.
71. In the first three years of Alessandri's presidency, the cost of living rose by 61%. See J. Petras, *Politics and Social Forces* . . . , op. cit., 64 above, p. 101.
72. See, J. Petras, 'Christian Democracy in Chile', op. cit., 64 above, p. 57.
73. See I. Roxborough, et al., *Chile: The State and Revolution*, op. cit., 64 above, p. 61.
74. On the increasing numbers and political significance of the urban white-collar stratum during this period, see I. Roxborough et al., *Chile: The State and Revolution*, op. cit., 64 above, Chapters 1 and 2. Urban 'employees' (as

opposed to urban workers) constituted 18.1% of the population in Chile in 1964, according to a study in 1964 by OCEPLAN, cited by A. G. Frank in his 'Capitalism and Underdevelopment in Chile', op. cit., 3 above.

75. See, notably, P. P. Rey, in his introduction to *Colonialisme, néo-colonialisme* . . . , op. cit., 19 above, pp. 11–24, where he gives a critique of the limitations of the Sartrean discourse from within which the key concepts of these texts are founded.
76. Recall that we specified the concept of dislocation in an earlier section of this chapter.
77. See Chapters 8 and 9.
78. On the relative autonomy of ideology within the social formation, see Chapter 5, pp. 118–19.
79. Such ideological combinations appear, for example, to have been crucial in the development of the Indonesian and Indian nationalist movements. On the Indonesian example, see B. O'G. Anderson, *Java in a time of Revolution*, Cornell University Press, Ithaca, 1972, particularly, Chapter 1.

Bibliography

L. Althusser, E. Balibar, R. Establet, J. Rancière, *Lire Le Capital*, (4 vols), Maspero, Paris, 1973.

L. Althusser, E. Balibar, *Reading Capital*, New Left Books, 1970.

L. Althusser, *For Marx*, Allen Lane, London, 1969.

S. Amin, *Accumulation on a World Scale*, Monthly Review Press, New York, 1975.

S. Amin, *Neo-Colonialism in West Africa*, Penguin African Library, London, 1973.

S. Amin, *Unequal Development*, Harvester Press, Brighton, 1977.

B. O'G. Anderson, *Java in a Time of Revolution: Occupation and Resistance, 1944–1946*, Cornell University Press, Ithaca, 1972.

A. T. Aymonier, *Histoire de l'Ancien Cambodge*, Paris, 1920.

A. T. Aymonier, *La Cambodge*, Paris, 1900–4.

A. Badiou, 'Le (Re) Commencement du Materialisme Dialectique', *Critique*, Paris, 1967.

P. Bairoch, *The Economic Development of the Third World since 1900*, Methuen, London, 1975.

J. Banaji, 'Backward Capitalism, Primitive Accumulation and Modes of Production', *Journal of Contemporary Asia*, Vol. 3, No. 4, 1973.

P. Baran, *The Political Economy of Growth*, Monthly Review Press, New York, 1957.

P. Baran and P. Sweezy, *Monopoly Capital*, Penguin, London, 1966.

M. Barratt-Brown, *The Economics of Imperialism*, Penguin, London, 1974.

C. Baudelot, R. Establet, J. Malemort, *La petite bourgeoisie en France*, Maspero, Paris, 1974.

G. Beckford, *Persistent Poverty*, Oxford University Press, London, 1972.

R. Bellah, *Tokugawa Religion*, Free Press, Chicago, 1957.

R. Bellah, 'Reflections on the Protestant ethnic analogy in Asia', *Journal of Social Issues*, No. 19, pp. 52–60.

R. Bendix, 'Tradition and Modernity Re-considered' in *Comparative Studies in Society and History*, No. 9, pp. 292–346.

L. Best, 'Outline of a Model of a Pure Plantation Economy', *Social and Economic Studies*, September 1968.
C. Bettelheim, *Economic Calculation and Forms of Property*, Monthly Review Press, New York, 1975, and Routledge and Kegan Paul, London, 1976.
C. Bettelheim, *The Transition to Socialist Economy*, Harvester Press, Brighton, 1975.
C. Bettelheim, *India Independent*, Monthly Review Press, New York, 1968.
C. Bettelheim, 'Theoretical Comments' in Arghiri Emmanuel, ed., *Unequal Exchange: A Study of the Imperialism of Trade*, New Left Books, London, 1972, Appendix I.
A. K. Bhattacharya, *Foreign Trade and International Development*, Teakfield, 1977.
M. Black, *The Social Theories of Talcott Parsons*, Prentice-Hall, Englewood Cliffs, New Jersey, 1961.
P. Bohannan and G. Dalton, *Markets in Africa*, Northwestern University Press, 1962.
R. Brenner, 'The Origins of Capitalist Development: a critique of neo-Smithian Marxism', *New Left Review*, 104, July–August, 1977.
L. P. Briggs, *The Ancient Khmer Empire*, Philadelphia, 1951.
G. Burchell, 'Discourse: terminable and interminable', *Radical Philosophy*, No. 18, Autumn 1977.
J. Buttinger, *The Smaller Dragon*, Praeger, New York, 1958.
L. Cadière, *Résumé de l'histoire d'Annam*, Quihnon, Paris, 1911.
F. Cayrac-Blanchard, *Le Parti Communiste Indonésien*, Armand Colin, Paris, 1973.
Centre d'Études et de Recherches Marxistes, *Sur le mode de production Asiatique*, Éditions Sociales, Paris, 1969.
Centre d'Études et de Recherches Marxistes, *Sur les Sociétés Pré-capitalistes*, Éditions Sociales, Paris, 1970.
A. Césaire, *Discours sur le colonialisme*, Présence Africaine, Paris, 1962.
J. Chesnaux, *Contribution à l'histoire de la nation Vietnamienne*, Paris, Éditions Sociales, 1955.
J. Chesnaux, *Le Vietnam, études de politique et d'histoire*, Maspero, Paris, 1968.
G. Coèdes, *The Indianised States of S.E. Asia*, East-West Centre Press, Honolulu, 1968.
A. Cutler and J. Taylor, 'Theoretical Comments on the Transition from Feudalism to Capitalism', *Theoretical Practice*, No. 6, May 1972.
R. Dahrendorf, *Class and Class Conflict in Industrial Society*, Routledge

and Kegan Paul, London, 1957.

R. Dahrendorf, 'Out of Utopia', *American Journal of Sociology*, No. 64, pp. 115–27.

A. Dauphin-Meunier, *Histoire du Cambodge*, Paris, 1961.

B. Davey, 'Modes of Production and Socio-Economic Formations', *South Asia Marxist Review*, Vol. I, No. 2, 1975.

T. Dos Santos, 'The The Structure of Dependence' in *American Economic Review*, May, 1970.

Yves Duroux, 'Theoretical Comments' in *Sur L'Articulation des Modes de production*, Cahiers de Planification, Pub. by École Pratique des Hautes Études, Centre d'Études de Planification Socialiste, Paris, 1970, Vol. 2, pp. 163–73.

S. N. Eisenstadt, 'Social Change, Differentiation and Evolution', *American Sociological Review*, No. 29, pp. 375–86.

S. N. Eisenstadt, *Modernisation, Protest and Change*, Prentice-Hall, Englewood Cliffs, New Jeresey, 1965.

S. N. Eisenstadt, *Tradition, Change and Modernity*, John Wiley, London, 1973.

Arghiri Emmanuel, 'Current Myths of Development', *New Left Review*, No. 85, May–June, 1974.

Arghiri Emmanuel, *Unequal Exchange: A Study of the Imperialism of Trade*, New Left Books, London, 1972.

F. Engels, *The Origin of the Family, Private Property and the State*, in K. Marx and F. Engels, *Selected Works*, Volume 2, Progress Publishers, Moscow, 1962, pp. 170–327.

F. Engels, *Anti-Dühring*, Progress Publishers, Moscow, 1969.

F. Fanon, *The Wretched of the Earth*, Penguin Books, London, 1967.

F. Fanon, *Black Skin, White Masks*, MacGibbon and Kee, London, 1968.

Fei Ling, *Proletarian Culture in China*, AREAS, London, 1973.

D. K. Fieldhouse, *The Theory of Capitalist Imperialism*, Longman, London, 1969.

D. Foss, 'The World View of Talcott Parsons', in M. Stein and A. Vidich, eds., *Sociology on Trial*, Prentice-Hall, Englewood Cliff, New Jersey, 1963.

M. Foucault, *The Order of Things*, Tavistock Press, London, 1970.

M. Foucault, *Madness and Civilisation*, Tavistock Press, London, 1965.

M. Foucault, *The Archaeology of Knowledge*, Tavistock Press, London, 1972.

M. Foucault, *The Birth of the Clinic*, Tavistock Press, London, 1975.

A. G. Frank, *Capitalism and Underdevelopment in Latin America*,

Monthly Review Press, New York, 1967.

A. G. Frank, *Latin America: Underdevelopment or Revolution*, Monthly Review Press, New York, 1969.

A. G. Frank, *Lumpenbourgeoisie, Lumpendevelopment – Dependence, Class and Politics in Latin America*, Monthly Review Press, New York, 1972.

A. G. Frank, J. Cockcroft, D. Johnson, *Dependence and Underdevelopment: Latin America's Political Economy*, Doubleday, New York, 1972.

J. S. Furnivall, *Netherlands India: A Study of Plural Economy*, Macmillan, New York, 1974.

J. S. Furnivall, *Colonial Policy and Practice: A Comparative Study of Burma and Netherlands India*, Cambridge, 1948.

C. Furtado, *Development and Underdevelopment*, University of California Press, Berkeley, 1971.

C. Geertz, *Agricultural Involution*, University of California Press, 1963.

C. Geertz, *The Religion of Java*, The Free Press, New York, 1960.

C. Geertz, *Peddlers and Princes*, University of Chicago Press, 1963.

E. Genovese, *The World the Slaveholders Made*, Vintage Books, New York, 1969.

M. Godelier, *Rationality and Irrationality in Economics*, New Left Books, London, 1972.

M. Godelier, *Horizon, trajets Marxistes en anthropologie*, Maspero, Paris, 1973.

M. Godelier, 'System, Structure and Contradiction in Capital', *Socialist Register*, London, 1967, pp. 91–114.

M. Godelier, 'Modes de production, rapports de parenté et structures demographiques', *La Pensée*, Paris, 1973.

C. Gosselin, *L'Empire d'Annam*, Paris, 1957.

B. P. Groslier, *Angkor et le Cambodge au XVIe siècle d'après les sources portugaises et espagnoles*, Paris, 1968.

A. Guerrero, *Philippine Society and Revolution*, Ta Kung Pao, Hong Kong, 1971.

J. S. Gusfield, 'Tradition and Modernity: Misplaced polarities in the study of social change', *American Journal of Sociology*, Vol. 72, No. 4, pp. 351–62.

T. Hayter, *Aid as Imperialism*, Penguin, London, 1971.

C. Hempel, 'The Logic of Functional Analysis' in L. Gross, ed., *Symposium on Sociological Theory*, Evanston, 1959.

M. Herskovits, *Economic Anthropology*, Knopf, New York, 1951.

R. Hilferding, *Das Finanzkapital*, Wiener Volksbuchandlung, Wien, 1923.

C. Hill, *From Reformation to Revolution*, Penguin Books, London, 1969.

P. Q. Hirst, *Social Evolution and Sociological Categories*, Allen and Unwin, London, 1976.

B. Hindess and P. Q. Hirst, *Pre-Capitalist Modes of Production*, Routledge, London, 1975.

B. Hindess and P. Q. Hirst, *Mode of Production and Social Formation*, Macmillan, London, 1977.

E. Hobsbawm, 'The General Crisis of the European Economy in the Seventeenth Century', *Past and Present*, 5 and 6, 1964.

E. Hobsbawm, *Industry and Empire*, Penguin Books, London, 1969.

A. Hoogvelt, *The Sociology of Developing Societies*, Macmillan, London, 1976.

B. F. Hoselitz, *Sociological Factors in Economic Development*, Free Press, Chicago, 1960.

B. F. Hoselitz and W. E. Moore, *Industrialisation and Society*, Mouton, the Hague, 1963.

A. Hussain, 'Godelier's Rationality and Irrationality in Economics', *Theoretical Practice*, London, No. 7/8, 1973.

M. Hussein, *Class Conflict in Egypt*, Monthly Review Press, New York, 1973.

P. Jalée, *The Third World in World Economy*, Monthly Review Press, New York, 1969.

P. Jalée, *The Pillage of the Third World*, Monthly Review Press, New York, 1968.

K. Kautsky, *La Question Agraire*, Maspero, Paris, 1970.

G. Kay, *Development and Underdevelopment: A Marxist Analysis*, Macmillan, London, 1975.

C. D. Kepner, *Social Aspects of the Banana Industry*, Columbia University Press, New York, 1936.

L. Krader, *The Asiatic Mode of Production: Sources, Development and Change in the writings of Karl Marx*, Van Gorcum, Essen, 1976.

W. Kula, *Théorie économique du système féodal*, Mouton, Paris, 1970.

E. Laclau, 'Imperialism in Latin America', *New Left Review*, London, No. 67, May–June, 1971.

D. Landes, *The Unbound Prometheus*, Cambridge University Press, 1970.

O. Lange, *Political Economy*, Pergamon Press, Oxford, 1963, Volume I.

E. Leach, 'Hydraulic Society in Ceylon', *Past and Present*, No. 15.

V. I. Lenin, *The Development of Capitalism in Russia*, Progress Publishers, Moscow, 1967.

V. I. Lenin, *Imperialism, the Highest Stage of Capitalism*, Progress Publishers, Moscow, 1968.

V. I. Lenin, *The Nascent Trend of Imperialist Economism*, Progress Publishers, Moscow, 1969.

M. J. Levy, *Modernisation and the Structure of Societies*, Princeton University Press, 1966.

G. Lichtheim, 'Marx and the Asiatic Mode of Production', *St. Anthony's Papers*, No. 14, 1963.

D. Lockwood, 'Social integration and System Integration', *Explorations in Social Change*, G. K. Zollschan and W. Hirsch, R. K. P., London, 1964.

A. D. Lublinskaya, *French Absolutism: The Crucial Phase*, Cambridge University Press, 1968.

R. Luxemburg, *The Accumulation of Capital*, Routledge and Kegan Paul, London, 1963.

R. Luxemburg and M. Bukharin, *Imperialism and the Accumulation of Capital*, Allen Lane, London, 1972.

D. McClelland, *The Achieving Society*, Von Nostrand, Princeton, 1961.

D. McClelland, 'Motivational Patterns in S.E. Asia with Special Reference to the Chinese Case', *Journal of Social Issues*, Vol. 29, No. I, January, 1963.

R. McVey, Introduction to *Nationalism, Islam and Marxism*, by Sukarno, Ithaca, Cornell University Press, New York, 1969.

R. McVey, *Indonesia*, New Haven, Connecticut, 1963.

H. Magdoff, *The Age of Imperialism*, Monthly Review Press, New York, 1966.

B. Malinowski, *A Scientific Theory of Culture*, Oxford University Press, 1969.

Mao Tse-Tung, 'An Analysis of Classes in Chinese Society', in *Selected Works*, Foreign Languages Press, Peking, 1967, Vol. I.

K. Marx, *Capital*, Volumes 1–3, Progress Publishers, Moscow, 1965, 1966, 1967, respectively.

K. Marx, *Grundrisse*, Penguin Books, London, 1973.

K. Marx, *A Contribution to the Critique of Political Economy*, Laurence and Wishart, London, 1971.

K. Marx, *The Poverty of Philosophy*, Progress Publishers, Moscow, 1966.

K. Marx and F. Engels, *On Colonialism*, Progress Publishers, Moscow, 1968.

K. Marx and F. Engels, *Selected Correspondence*, Progress Publishers, Moscow, 1965.

K. Marx, 'The Class Struggles in France', and 'The Eighteenth Brumaire of Louis Bonaparte' in K. Marx and F. Engels, *Selected Works*, Progress Publishers, Moscow, 1962, Volume I.

C. B. Maybon, *Histoire Moderne du Pays d'Annam*, (1592–1820), Paris, 1920.

C. Meillassoux, *Anthropologie économique des Guoro de Côte d'Ivoire*, Mouton et Cie, Paris, 1964.

C. Meillassoux, 'Essai d'interpretation des phénomènes économiques dans les sociétés traditionelles d'autosubsistence', *Cahiers d'Études Africaines*, 1960, No. 4.

U. Melotti, *Marx and the Third World*, Macmillan, 1977.

A. Migot, *Les Khmer, des origines d'Angkor au Cambodge d'aujourd'hui*, Paris, 1960.

M. Molyneux, 'Androcentrism in Marxist Anthropology', *Critique of Anthropology*, Vols. 9/10, 1977.

R. Mortimer, 'Class, Social Cleavage and Indonesian Communism', *Indonesia*, 8 October 1969, pp. 1–20.

R. Mortimer, ed., *Showcase State: The Illusions of Indonesia's 'accelerated modernisation'*, Angus and Robertson, London and Sydney, 1973.

R. Mortimer, *Indonesian Communism under Sukarno*, Cornell University Press, Ithaca, 1974.

R. A. Nisbet, *Social Change and History*, Oxford University Press, 1969.

R. Owen and B. Sutcliffe, *Studies in the Theory of Imperialism*, Longman, London, 1972.

I. Oxaal, T. Barnett, D. Booth, eds., *Beyond the Sociology of Development*, Routledge and Kegan Paul, London, 1975.

T. Parsons, *The Social System*, Routledge and Kegan Paul, London, 1971.

T. Parsons, *The Structure of Social Action*, McGraw-Hill, New York, 1948.

T. Parsons, E. Shils, et al., eds., *Theories of Society*, Free Press, Glencoe, New York, 1961.

T. Parsons, *Essays in Sociological Theory*, Free Press, Glencoe, New York, 1954.

T. Parsons, *The Structure of Social Action*, McGraw-Hill, New York, 1948 (two volumes).

T. Parsons (with E. Shils), *Towards a General Theory of Action*, Harvard Press, Cambridge, Mass., 1951.

T. Parsons, 'The Pattern Variables re-visited: A Response to Robert Dubin', *American Sociological Review*, Vol. 25, No. 4, pp. 192–219.

T. Parsons, *Societies: Evolutionary and Comparative Perspectives*, Prentice-Hall, New York, 1966.

T. Parsons, *Structure and Process in Modern Societies*, Free Press, Glencoe, New York, 1960.

T. Parsons, *Sociological Theory and Modern Society*, Free Press, New York, 1960.

C. Payer, *The Debt Trap: The IMF and the Third World*, Penguin Books, London, 1974.

J. Petras, *Politics and Social Forces in Chilean Development*, University of California Press, Berkeley, 1969.

J. Petras, ed., *Latin America: From Dependence to Revolution*, New York, 1975.

J. Petras, P. McMichael, R. Rhodes, 'Industry in the Third World', *New Left Review*, No. 85, May–June, 1974.

A. Phillips, 'The Concept of Development', *Review of African Political Economy*, No. 8, January–April, 1977.

T. G. Pigeaud, *Java in the Fourteenth Century: A Study in Cultural History*, The Hague, 1960–3 (5 volumes).

G. Pilling, 'Imperialism, trade and "unequal exchange", the work of Arghiri Emmanuel', *Economy and Society*, Vol. 2, No. 2, pp. 164–86.

J. Pluvier, *Confrontations: A Study in Indonesian Politics*, Oxford University Press, Kuala Lumpur, 1965.

K. Polanyi, *Trade and Markets in the Early Empires*, The Free Press, Glencoe, 1957.

K. Polanyi, *Dahomey and the Slave Trade*, A. E. S. Monograph, Number 42, University of Washington Press, 1966.

K. Polanyi, *Primitive, Archaic and Modern Economics*, Doubleday, New York, 1968.

N. Poulantzas, *Classes in Contemporary Capitalism*, New Left Books, London, 1975.

Vu Quoc Thue, *L'Économie communaliste du Vietnam*, ed., De Droit, Paris, 1951.

A. R. Radcliffe-Brown and D. Forde, *African Systems of Kinship and Marriage*, Oxford University Press, 1950.

T. S. Raffles, *History of Java*, London, 1817.

J. Rex, *Key Problems in Sociological Theory*, R.K.P., London, 1961.

P. P. Rey, *Colonialisme, néo-colonialisme et transition au capitalisme*, Maspero, Paris, 1971.

P. P. Rey, *Les alliances de classes*, Maspero, Paris, 1973.

P. P. Rey and G. Dupré, 'Reflections on the pertinence of a theory of the History of Exchange', *Economy and Society*, Vol. 2, No. 2.

R. I. Rhodes, ed., *Imperialism and Underdevelopment: A Reader*, Monthly Review Press, New York, 1970.

D. Ricardo, *The Principles of Political Economy and Taxation*, ed., R.M. Hartwell, Penguin Books, London, 1971.

J. Riche, 'La Cochinchine au XVIIIe siècle', *Revue d'Europe et des Colonies*, Paris, 1906.

L. Robbins, *An Essay on the Nature and Significance of Economic Science*, London, 1932.

C. Robequain, *The Economic Development of French Indo-China*, Oxford University Press, London, 1944.

G. Rocher, *Talcott Parsons and American Sociology*, Nelson, London, 1974.

R. Rosdolsky, *Zür Entstehungsgeschichte des Marxischen "Kapital"*, Europäische Verlagsanstalt, Wien, Frankfurt-am-Main, 1968 (2 volumes).

C. E. Rouger, *Histoire Militaire et Politique de L'Annam et Tonkin depuis 1799*, Paris, 1906.

I. Roxborough, P. O'Brien and J. Roddick, *Chile: The State and Revolution*, Macmillan, London, 1977.

I. Roxborough, 'Dependency Theory in the Sociology of Development: Theoretical Problems', *West African Journal of Sociology and Political Science*, Vol. I, No. 2, January, 1976.

Gilles Sautter, 'Notes sur la construction du chemin de fer Congo-Océan, (1921–34)', *Cahiers d'Études Africaines*, No. 26, 1967, pp. 219–99.

F. M. Schnitger, *Forgotten Kingdoms in Sumatra*, E. J. Brill, Leiden, 1939.

A. D. Smith, *The Concept of Social Change*, Routledge and Kegan Paul, London, 1973.

T. C. Smith, *Agrarian Origins of Modern Japan*, Stanford University Press, 1959.

G. Sofri, *Il modo di produzioni asiatico*, Einaudi, Turino, 1963.

B. Sutcliffe, *Industry and Underdevelopment*, Addison-Worsley, London, 1971.

P. Sweezy, *The Theory of Capitalist Development*, Monthly Review Press, New York, 1942.

J. G. Taylor, 'The Economic Strategy of the New Order', British Indonesia Committee, ed., *Repression and Exploitation in Indonesia.*

J. G. Taylor, 'New-Marxism and Underdevelopment—A Sociological Phantasy', *The Journal of Contemporary Asia*, Vol. 4, No. 1.

J. G. Taylor, 'Pre-Capitalist Modes of Production', *Critique of Anthropology*, Nos. 4 and 5/6 Autumn 1975 and Spring 1976.

E. Terray, *Marxism and 'Primitive' Societies*, Monthly Review Press, New York, 1972.

E. Terray, 'Long-distance exchange and the formation of the state: the case of the Abron Kingdom of Gyaman', *Economy and Society*, Vol. 3, No. 3.

E. Terray, 'Classes and Class Consciousness in the Abron Kingdom of Gyaman', in M. Bloch, ed., *Marxist Analyses and Social Anthropology*, Association of Social Anthropologists of the Commonwealth (ASA) Publications, London, 1975.

E. Terray, *L'Organisation Sociale des Dida du Côte d'Ivoire*, Annales de l'Universite d'Abidjan, Serie F, Vol. 1, fascule 2, 1969.

Le Thanh-Khoi, *Le Vietnam Histoire et Civilisation*, Éditions de Minuit, Paris, 1955.

Nguyen Thanh-Nha, *Tableau économique de Viètnam au XVIIe et XVIIIe siècles*, Éditions Cujas, Paris, 1970.

E. T. Thompson, 'The Plantation as a Social System' in *Plantation Systems of the New World, ed., by Pan American Union.*

M. P. Todaro, *Economic Development in the Third World*, Longman, London, 1977.

United Nations Publications:

The U.N. Statistical Yearbook (published annually).

The U.N. Yearbook of International Trade Statistics.

The Yearbook of Labour Statistics, published by the International Labour Office (Geneva).

The Production Yearbook, published by the Food and Agricultural Organisation of the United Nations (Rome).

The Yearbook of National Accounts Statistics.

The Balance of Payments Yearbook, published by the International Monetary Fund (New York).

E. Utrecht, 'Land Reform and Bimas in Indonesia', *Journal of Contemporary Asia*, Vol. 3, No. 2.

E. Utrecht, 'Land Reform in Indonesia', *Bulletin of Indonesian Economic Studies*, November 1969.

R. Van Heine-Gelden, *Conceptions of State and Kingship in S.E. Asia*, Ithaca, New York, 1963.

Nguyen Van Huyen, *La civilisation annamite*, Hanoi, 1944.

C. B. Walker, *Angkor Empire*, Calcutta, 1955.

I. Wallerstein, *The Origins of the Modern World System*, New York, 1974.

I. Wallerstein, 'Dependence in an Interdependent World', *African Studies Review*, 17, April, 1974.

B. Warren, 'Myths of Underdevelopment', in *New Left Review*, No. 81, May–June, 1974.

W. F. Wertheim, *Indonesian Society in Transition*, Van Hoeve, The Hague, 1969.

P. Win and C. Kay, 'Agrarian Reform and Rural Revolution in Allende's Chile', *Journal of Latin American Studies*, Vol. 6, 1974.

K. Wittfogel, *Oriental Despotism*, Yale University Press, 1957.

C. Wright-Mills, *The Sociological Imagination*, Oxford University Press, New York, 1959.

O. W. Woters, *Early Indonesian Commerce: A Study of the Origins of Srivijaya*, Ithaca, New York, 1967.

A. Zammit, ed., *The Chilean Road to Socialism*, Institute of Development Studies, University of Sussex, Brighton, 1973.

Index

Superior numbers indicate note numbers.